KEY NOTES ON
ANIMAL SCIENCE AND DAIRY SCIENCE

For Ready Reference to the

**STUDENTS, TEACHERS, RESEARCHERS & ASPIRANTS OF COMPETITIVE
EXAMINATIONS**

THE EDITORS

Dr. U.D. Chavan obtained his M.Sc. (Agri. in Biochemistry) degree from Mahatma Phule Krishi Vidyapeeth, Rahuri. He received his Ph.D. degree in Food Science from Memorial University of Newfoundland St. John's Canada in 1999. He has done International Training on "Global Nutrition 2002" at Uppsala University Uppasala, Sweden in 2002. Dr. Chavan worked as Senior Research Assistant in the Department of Biochemistry & Food Science and Technology at MPKV Rahuri from 1988 to 2000. During his Ph.D., he worked as Technician/Research Associate at Atlantic Cool Climate Crop Research Center and Agriculture and Agri-Food Canada. He received D.Sc. degree in 2006 from USA.

Dr. Chavan is presently working as a Senior Cereal Food Technologist in the Department of Food Science & Technology at Mahatma Phule Krishi Vidyapeeth, Rahuri.

Dr. J.V. Patil obtained his M.Sc. (Agri.) from, MPKV, Rahuri. He completed his course work for Ph.D. at CCSHAU, Hisar and research at MPKV, Rahuri in 1992. He rendered his research and teaching services at MPKV Rahuri as Geneticist, Associate Professor, Plant Breeder and Professor of Genetics & Plant Breeding and Head, Genetics and Plant Breeding Department, MPKV, Rahuri. He also delivered many administrative responsibilities in the University. Dr. Patil joined as the Director, Directorate of Sorghum Research, Hyderabad in August 2010.

THE CONTRIBUTORS

Dr. K.D. Chavan is an Assistant Professor (Animal Science & Dairy Science) in College of Agricultural, Pune at Mahatma Phule Krishi Vidyapeeth, Rahuri.

Dr. S. Mandakmale is an Associate Professor in the Department of Animal Science and Dairy Science at Mahatma Phule Krishi Vidyapeeth, Rahuri.

KEY NOTES ON
ANIMAL SCIENCE AND DAIRY SCIENCE

For Ready Reference to the

STUDENTS, TEACHERS, RESEARCHERS & ASPIRANTS OF COMPETITIVE EXAMINATIONS

Editors

U.D. CHAVAN
&
J.V. PATIL

Contributors

K.D. CHAVAN
S. MANDAKMALE

2015
Daya Publishing House®
A Division of
Astral International (P) Ltd
New Delhi 110 002

© 2015 PUBLISHER
ISBN: 9789351307013 (International Edition)

Published by   : **Daya Publishing House®**
                           A Division of
                           Astral International Pvt. Ltd.
                           – ISO 9001:2008 Certified Company –
                           4760-61/23, Ansari Road, Darya Ganj
                           New Delhi-110 002
                           Ph. 011-43549197, 23278134
                           E-mail: info@astralint.com
                           Website: www.astralint.com

Laser Typesetting   : **Twinkle Graphics, Delhi**

Printed at   : **Thomson Press India Limited**

PRINTED IN INDIA

PREFACE

India is an agricultural country. The Indian economy is basically agarian. Inspite of economic and industrialization, agriculture is the backbone of the Indian economy. As Mahatma Gandhi said "India's lives in villages and agriculture is the soul of Indian economy". Agriculture is a vast subject and encompasses at least 20 major and minor subjects in it. New developments have lead to entirely a new face of agriculture. Study of agriculture has always been intrigued with a mosaic of interwove concepts, subjects, facts and figures. There are number of books and large literature on Animal Science and Dairy Science but the Key Notes type of book have not been compiled in a readable manner.

The present book *"Key Notes on Animal Science and Dairy Science"* has been designed to fulfill this long felt need of students, teachers, researchers and aspirants of competitive examinations. It is designed in such a way that give rapid, easy access to the core materials in a short format which facilitates easily learning and rapid revision. The book carries fundamentals of Animal Science and Dairy Science. There are six chapters elaborating Discoveries, Abbreviations, Terminology, Short Explanations, Reasoning, Distinguish between as well as references also included. The most recent information is provided along with a detailed list of references for further reading.

Hope this book would be highly useful for graduate and post-graduate students of agriculture, teachers and researchers. This book will also useful for the aspirants of various competitive examinations such as Agricultural Research Service (ARS), ICAR- National Eligibility Test (NET), State Eligibility Test (SET), Junior Research Fellowship (JRF), Senior Research Fellowship (SRF), Civil Services, Allied Agricultural Examinations and Extension Workers for reference and easy answers of many complicated questions. Thus it is expected that this book will adequately meet the need of wider circle of students and readers for preparing their professional career.

We acknowledge the references that are used in this manuscript. Authors are also thankful to all scientists and friends who have helped directly or indirectly while preparing this manuscript. The editors of grateful to all the contributors for their cooperation, support and timely submission of their manuscripts for bringing out this publication. We would have like to acknowledge the patience and support of our families whilst we have spent many hours with drafts of manuscripts rather than with them. Lastly, our sincere thanks to publisher Astral International Pvt. Ltd., New Delhi who provides an opportunity to publish this book.

To all readers we extend an invitation to report that no doubts have escaped our attention and to offer suggestion for improvements that can be incorporated in future editions.

U.D.Chavan and J.V. Patil

Editors

CONTENTS

1

DISCOVERIES

Scientist	Year	Discovery
G.J. Mulder	-	Gave name protein to Nitrogenous food. Protein : to take first place
Essher and Kauffmann	-	Gelatin was deficient in tyrosine and Cystine
Leng and Nolan	-	Used isotope dilution and stable isotope provided detailed information on VFA production and N transactions in recital rumen
Kurt Nehring	-	Established "Oscar Kellner institute for Animal nutrition" in R ostock. Worked out Rostock system of feed evaluation
G.P. Lofgreen	-	Net energy system in the evaluation of feed staff and energy requirement of livestock
W.N. Garrett	-	Net energy tables for use in feeding beef cattle
J.P. Fontenot	-	Magnesium requirement in cattle and sheep and it metabolism Munities valuate broiler littler for ruminants
Peter J. Van Spest	-	Procedure for fiber estimation Developed detergent system for fiver estimation Discovered maillaved reaction
K.C. Sen, S.N. Ray, and S.K. Ranjhan	-	Nutritive value of Indian cattle feeds and feeding of farm animals
N.D. Kehar	-	Nutritive value of non conception feeds. Enrichment of straw by wet alkali treatment
R.E. Hungate	-	Rumen protozoa
R.T. Holman	-	Omega system for essential fatty acids
Pasture and Robert Koch	-	Anthrax in sheep caused by bacterium
Joseph lister	-	Antiseptic sagery
Robert Hooke	-	Applied term cell to cavities in cork
Marcello Malpighi	-	Published anatomy of plant
Santario Sanctrius	1561-1636	First experiment of metabolism. Weighed him of before and after food intake.
Zacharias Jansen	1590	Compound microscope
Van leeuenhock	1675	Microscope
Linnaeus	1707-1998	Taxonomy a Science of Classification
Lazaro Spallanzani	1729-1799	Swallowed linen bags containing meat and bread and retrieved by strings attached to them and found chemical changes

Scientist	Year	Discovery
Antoine Laurent Lavoisier	1743–1794	Father of Nutrition used balance and thevemoete in nutrition studies. Designed calorimeter. Discovered that combustion was an oxidation and respiration in body involved the combination of carbon and hydrogen depend on food intake and work done
Lazzaro spallanzani	1777	Successful AZ in reptiles
Francois Magendie	1783-1855	French physiologist N Present in body had its origin in N compounds present in food. Published Gelatin report in 1841 stated that all proteins not of equal value
Justus von Liebig	1803-1873	Nitrogenous compounds were utilized for body building and nonprotein to produce heat
John B. Laues and Joseph lt. Gilbert	1814-1900 1871-1901	Pioneer and Agronomy and Animal nutrition analyzed bodies of farm animal
Thedor Schwann	1839	Formulated cell theory
Gruby and Delafond	1843	Rumen Protozoa
Stephen M. Babcock	1843-1931	Babcock test: Purified died method resulted in discovery of first vitamin in 1913
Henry Prenfiss Armsby	1851-1921	Constructed respiration calorimeter for farm animals and studied Heat production in cattle, which let to net energy system for feed evaluation
Max Rubner	1854-1932	Energy Metabolism Carbohydrates and fat. Were interchangeable in metabolism on the basis of energy equivalent
Max Rubner and E.F. Adolph	1854-1932 -1933	Vital Role of water in body
Thomas B. Osborne and Lafayette B. Mendei	1859- 1929 1872-1935	Discoveries in protein & vitamins
Wilhelm Henneberg and Friedrich stohmann	1865	Proximate analysis
Tappeiner	1884	Large quality of VFA notably acetic acid produced from in-vitro fermentation of cellulose by bacteria.
Colin	1886	Permanent Rumen Fistula technique.
F.B. Morrison	1887-1958	Feeds and feeding standards
Leonard Amby Maynard	1887-1972	NRC feeding standards nutrient allowance for various animals
Max Kleiber	1893-1976	Developed the use of 0.75 powers instead of surface area to describe energy metabolism.
R.W. Swift	1895-1975	Energy metabolism with Arms by calorimeter.
Elias I Ibanoft	1899	AZ in ford, Horses cattle, Sheep
Burch H. Scheider	1901-73	Feeds of the world : their digestibility and composition. Evaluation of feed through digestibility experiments.
E.J. Underwood	1905-80	Trace Elements in human and Animal Nutrition.
Hart et al	1928	Necessary of copper for hemoglobin formation.

Scientist	Year	Discovery
W.C. Rose	1930	Defined EAA and Classified 10 Amino acids as essential and other non essential
Franke and Potter	1935	Selenium as toxic facts in forage. (Alkali discard, blind staggers in America sulimana in Mexican and Degnala disease in India are in same group)
Sampath Kumar	1939	First Attempted AI in India.
Wilbur Olin Atwater and E.B. Rosa	1944-07	Energy requirement for various body functions and nutritive value of food.
Nathan Zuntz	1947-1920	Pioneer in basal metabolism. First portable respiration apparatus. In 1879 formulated fermentation hypothesis in ruminants.
Erle Bartey	1949	Development of Polox alene to prevent Bloat
Oscar Kellner	1951-11	Practical application of feeding standards, starch equivalent system of energy evaluation.
Leg and Annison	1964	Ketone body metabolism in whole animal.
Asoknath Bhattacharya	1964	Supplementation of dried beet pulp, citrus pulp as a grain replacement.
Tony Joseph Cunha	1965	Nutrient requirement in swine
Carlisle	1970	Silicon is essential for normal calcification of chick bone.
C.G. Orpin and T. Bauchop	1975	Rumen Fungi
Scott and Cook	1975	By pass lipid and protein.
H.H. Mitchell	1986-66	Comparative nutrition of man and DomesticAnimals
Nielsen	1990	Role of vanadium for regulation of Nat, Kt, ATP's protein kinase

2

ABBREVIATIONS

Abbreviation	Full Form
ADF	Acid detergent fiber
ACTH	Adrenocortico tropic hormone
ADF	Acid Detergent Fiber
ADH	Artidiuretic hormone
ADI	Acceptable daily intake
ADIN	Acid Detergent Insoluble Nitrogen
ADL	Acid Detergent Lignin
ADSA	Score card of table butter
AFD	Accelerated freeze drying
AI	Artificial Insemination
AIA	Acid Insoluble Ash
AMZ	Australian milking Zebu
AOAC	Association of Official Analytical Chemists
AV	Artificial vagina
BCR	Benefit cost ratio
BHA	Butylated hydroxyl anisole
BIS	Bureau of Indian Standard
BMR	Basal metabolic rate
BQ	Black Quarter
BR	Butyro-Refractometer
BV	Biological Value
CA	Cream acidity
CAC	Codex alimentarius commission
CAO	Calcium oxide

Abbreviation	Full Form
CAR	Corrective action report
CAW	Citric acid whey
CD	Critical difference
CD	Coefficient of digestibility
CF	Crude fiber
CFR	Code of federal regulations
CFU	Colony forming unit
CIP	Cleaning in place
CL	Corpus luteum
CLA	Conjugated linoleinic acid
CMC	Carboxy methyl cellulose
COB	Clot-on-boiling test
CP	Crude protein
CRF	Critical recovery factor
CSA	Cream serum acidity
CT	Curd tension
CTAB	Cetyl trimethyl ammonium bromide
CUE	Cornell University Extender
DC	Digestion coefficient
DCP	Digestible crude protein
DE	Digestible energy
DFM	Direct fed microbial (term used for probiotics in US)
DHA	Docosa Hexaenoic Acid
DHEA	Dehydroepian drosterone
DM	Dry Matter
DMC	Direct Microscopic Count
DRI	Dairy Research Institute (New Zealand)
EAA	Essential Amino Acid
EB	Energy balance
EBT	Erichrome black T indicator

Abbreviation	Full Form
EBW	Empty Body weight
EDTA	Ethylene diamine-tetra acetic acid
EE	Ether extract
EFA	Essential Fatty Acid
EIC	Export Inspection Council of India
EMP	Enborn Meyerhof Pathway
EPA	Eicosa Pentaenoic Acid
ER	Endoplasmic reticulum
ET	Embryo transfer
EUN	Endogenous Urinary Nitrogen
EYC	Egg yolk citrate
EYGB	Egg Yolk glucose Sodium bicarbonate
EYGC	Egg Yolk Glucose Sodium Citrate diluents
EYP	Egg Yolk phosphate
FCBT	Float controlled balance tank
FCR	Feed conversion Ratio
FDV	Flow diversion value
FE	Fecal energy
FER	Feed efficiency Ratio
FMD	Foot mouth Disease
FSH	Follicle Stimulating hormone
FSMSC	Food safety management systems-based certification
GATT	General agreement on tariffs and trade
GE	Gross Energy
GH	Growth hormone
GMP	Good manufacturing practices
GNC	Groundnut cake
GPD	Gaseous product of digestion
GTH	Gonadotrophic hormone
H.C.	Hemicelluloses

Abbreviation	Full Form
HI	Heat increment
H^2	Heritability
HBPM	Hatchery By-Product-Meal
HC	Hemicelluloses
HCN	Hydrocyanic acid
HI	Heat Increment
HMS	Hexose monophosphate Shunt
HS	Hemorrhagic Septicemia
HTST	High temperature short time
IBR	Infections Bovine Rhinotracheitis
ICMR	Indian council of medical research
ICSH	Interstitial cell stimulating hormone
IDP	Inter departmental panel
IPC	In place cleaning
IPQC	In-process quality control
ISO	International Standardization organization
IU	International Units
IUPAC	International Union of physical and Analytical Chemists
IVDMD	*In Vitro* Dry Matter Digestibility
IVT	Illini Variable Temperature
J	Joule
KJ	Kilo Joule
LAB	Lactic acid bacteria
LH	Luteinizing hormone
LN_2	Liquid Nitrogen
LMPC	Low mucilage protein concentrates
LMPI	Low mucilage protein isolates
LNA	Linolenic Acid
LP	Lactoperoxidage
LTH	Luteotrophic hormone

Abbreviation	Full Form
LTLT	Low temperature long time
LTLT	Long time low Temperature
MAP	Modified atmospheric packaging
MAS	Marker assisted Selection
MBR	Methylene blue reduction
ME	Metabolisable energy
MFN	Metabolic Faucal Nitrogen
MJ	Mega Joule
MMPO	Milk and Milk Product Order
MMR	Methylene Blue Reduction
MOET	Multiple ovulation and embryo transfer
MRLs	Maximum residues limits
MSH	Melanocyte Stimulating hormone
MSNF	Milk solids not fat
MT	Metric tonnes
MYPDCI	Milk yield per day of calving interval
MYPDLL	Milk yield per day of lactation length
NFE	Nitrogen Free Extract
NDF	Neutral detergent fiber
NE	Net energy
ND	Neutral Detergent
NDDB	National dairy development board
NDF	Neutral Detergent Fiber
NDGA	Nordihydroguiaretic acid
NDRI	National Dairy Research Institute
NDS	Neutral Detergent Soluble
NE	Net Energy
NFDM	Non-fat-dry-milk
NFE	Nitrogen free extract
NIRD	National Institute for Research and Development (UK)

Abbreviation	Full Form
NK	Natural killer
NMR	Nuclear magnetic Resonance
NPN	Non protein nitrogenous compound
NPU	Net Protein Utilization
NR	Nutritive ratio
NRC	National Research Council
NRS	Non Reducing Sugars
NTRS	Near Infrared Spectroscopy
OR	Oxidation reduction
PER	Protein efficiency ratio
PFA	Prevention of food adulteration
PIF	Prolactin inhibitory factor
PMO	Pasteurized milk ordinance
PRV	Protein replacement value
RBC	Red blood cells
RBD	Randomized block design
RCBD	Randomized complete block design
RDA	Recommended dietary allowances
REF	Renal Erythropoetic factor
RF	Releasing factor
RM	Reichert Meissl
RS	Reducing Sugars
SDA	Specific Dynamic Action
SDE	Specific Dynamic Effect
SE	Standard error
SMP	Skim milk powder
SNF	Solid not fat
SO	Super Ovulation
SPS	Sanitary and phytosanitary
SRL	Strained Rumen Liquor
SSHE	Scraped surface heat exchanger
SSOPs	Sanitation standard operating procedures

Abbreviation	Full Form
STH	Somatotrophin hormone
TA	Total Ash
TB	Tuberculosis
TBG	Thyroxin binding globulin
TBPA	Thyroxin binding pre albumin
TBT	Technical barriers to trade
TCA	Tricarboxylic Acid
TCRPV	Tissue Cultural Rinderpest Vaccine
TCT	Thyro calcitocin
TDN	Total digestible nutrients
TEA	Triethanolamine
TPP	Thiamine Pyrophosphate
TRF	Thyroid releasing factor
TS	Total sugars
TSH	Thyroid stimulating hormone
TVFA	Total volatile fatty acids
UDP	Uridine Diphosphate
UDPG	Urdine Diphosphate Glucose
UE	Urinary energy
UFA	Unsterilized fatty acids
UHT	Ultra high temperature
UHTP	Ultra high temperature pasteurization
USDA	United States Department of Agriculture
USP	United States Pharmacpria
USPHS	United States public health sennices
Vitamin D_2	Ergocalciferol
Vitamin D_3	Cholecalciferol
Vitamin E	Tocopherol
Viz	*Videlicet* (namely)
WBC	White blood cells
WMP	Whole milk powder
WP	Wetable powder
WPN	Whey Protein Nitrogen
WTO	World Trade Organization

3

TERMINOLOGY

Term	Terminology
Allergic Diseases	Allergic reactions of animal body also cause some diseases, e.g., photosterilization, serum shock, allergic dermatitis, hay fever etc.
Abomasums	The fourth compartment of ruminant stomach (true stomach)
Absorption	The passage of material across a biological membrane
Acetonemia	Abnormally elevated concentration of ketone bodies in the body tissues and fluids.
Achromotricia	Loss of pigment in hair
Actin	Muscle fiber fraction that complexes with myosin to bring about muscle contraction
Adaptation	The adjustment of an organism to a new or changing environmental condition.
Additive	An ingredient or combination of ingredients in micro-quantities added to the basic feed mix or parts.
Additive Premix	(As defined by FDA) an article that be diluted for safe use in a feed additive. Concentrate, a feed additive supplement or a complete feed.
Adiabatic	No gain or loss of heat.
Adipose	Of a fatty nature.
Aerobic	A term usually applied to microorganisms that require oxygen to live and reproduce.
Aestivation (summer hibernation)	A dormant state in which some desert animals pass the hottest; time of summer.
Alopecia	Loss of hair.

Term	Terminology
Ambient Temperature	Temperature of fluid (usually air) that surrounds object on all sides.
Ammoniated	Impregnated with ammonia or an ammonium compound.
Amylase	Any one of several enzymes which aid in the hydrolysis of starch to maltose, for example, pancreatic amylase (amylopsin) and salivary amylase (ptyalin).
Anabolism	Synthesis of compounds.
Anaerobic	Oxygen not present.
Anatomy	Study of structure of body.
Anemic	Lacking in size and or number of red blood cells.
Aneurysm	Dilatation of the wall of artery.
Angelology	Study of circulatory system (heart and vessels).
Animal Husbandry	Science as well as an art of management including scientific feeding, breeding, housing, health cover of common domestic for maximum returns.
Animal Science	The science which deals with the feeding, breeding and management practices of the livestock.
Anorexia	Lack or loss of appetite.
Antibiotics	Group of soluble organic substances produced from microorganisms, which in small concentration inhibits the growth of other microbes.
Antibody	Substance produced in the body that acts against disease.
Antigen	A high-molecular-weight substance (usually protein) that, when foreign to the bloodstream of an animal, stimulates the formation of a specific antibody and reacts specifically *in vivo* or *in vitro* with its homologous antibody.
Antioxidant	Substance having the property of protecting other substance from oxidation such as vitamin E protects unsaturated fatty acids by oxidizing themselves.
Antioxidants	Ingredients which limits the oxidative spoilage.
Arthology	Study of Articular system (joints).

Term	Terminology
Artificial insemination	It is a technique in which semen with live sperms collected from the male and deposited in the female genetical tract by mechanical means.
Aspirated, Aspirating	(process) having removed chaff, dust or other light materials by use of air.
Atrophy	A wasting away of a part of the body.
Auto sexing	Sex differentiation at day old age some visual characters such as colour of down, early feathering etc.
Back fatter	A fat pig too heavy for bacon trade.
Bacon	Meat from back and sides of a pig.
Baconer	A bacon pig.
Balanced	The ratio, or feed having all known required nutrients in proper amount and proportion based upon recommendations of recognized authorities in the field of animal nutrition.
Balanced Ration	A ration which provide essential nutrients to the animal in such proportion and amounts that are required for the nourishment of the particular animal for 24 hrs.
Bangers	Cows that have reacted to Brucellosis test.
Barren sow	Sterile mature female pig.
Barrow	A mature pig whose testes were removed before reaching breeding age.
Basal Metabolism	Quantity of chemical energy expanded for body maintenance measured under specific conditions, e.g., standard metabolism, tissue activity of physiochemical changes of arresting animal.
Billy goat	Buck/male goat.
Biochemical oxygen demand (BOD)	The quantity of oxygen used by bacteria on the oxidation of organic matter in a specific time, at a specified temperature, and under specified conditions.
Biological value	The efficiency with which a protein furnishes the proper proportion and amounts of the essential amino acids. A protein which has a high biological value is said to be good quality.

Term	Terminology
Bitch	A female dog.
Blending (process)	To mingle or combine two or more ingredients or feeds.
Bloat	A disorder of ruminants caused by excess gas in the rumen.
Blood meal	Ground and dried meal of blood having CP ranges from 79-85%, used as protein rich ingredient in animal feed.
Blue pig	Pig produced by crossing a white breed with black breed of pigs.
Boar	An uncastrated adult male pig used for breeding purpose.
Bobby calves	Calves usually of only one week old or so.
Boiler Efficiency	The heat of condensation divided by the heat of combustion of the fuel used to produces the steam.
Bomb Calorimeter	An instrument used to determine the gross energy content of a material.
Bone meal row	Dried and ground bone meal obtained from un-decomposed bones with steams under pressure.
Bone meal steamed	The ground product sterilized by cooking un-decomposed bones with steams under pressure.
Bow Wow	A small, stunted aged steer with no quality.
Brahmini bull	The bull which is left in the name of dead person, a practice prevalent in certain areas.
Bran	The pericarp or seed coat.
Breed	A breed is a group of individuals of common ancestors which are genetically identifiable because all members of the genes present in one breed are also present at low frequency in all other breeds of the same species.
Breeding bull	A bull used for breeding purpose.
Breme	Female yak.
British Thermal Unit (BTU)	An engineering unit of heat energy equal to the quantity of heat necessary to raise the temperature of one pound of water one degree Fahrenheit.

Term	*Terminology*
Broiler	A chicken of either sex, usually of 8-10 weeks age that is tender meated flexible breast bone, cartilage, suitable for flying.
Brood mare	A mare kept for breeding.
Brooder /Hover	An apparatus used for keeping chickens warm.
Brooding	Rearing of chicks after hatching till are upto which warmth has to be provided
Buck	Sexually matured male goat used for breeding, Buckling: A male goat between 1 and 2 years age.
Buck-Billy	An uncastrated male used for breeding purpose.
Buffer	Any substance that can counteract changes in free acid or alkali concentration
Bull	An uncastrated male ox used for breeding purpose.
	An uncastrated sexually matured male.
Bull-Calf	A male calf under one year of age.
Buller	A cow apparently always in heat, oestrus.
Bullock or Steer	A castrated male ox of over 2 year's age.
Bullock/Steer	A castrated male ox of over two years.
Cad/Crittfng/Runt	Smallest pig of litter farrowed in the last, gunner/ rating runner-weaned pig.
Calf	A young one of cow class (ox) of either sex.
Calf	A young animal of bovine species under one year age.
Calving	An act of giving birth to young one by a female of cow class or Parturition in cattle.
Calving Interval	It is the period between the two successive calving.
Candling	Examination of eggs for finding egg objects such as cracks on shell, blood spots, loose cell etc. by holding the eggs between the eye and light source.
Canula	A tube for insertion into a body cavity.
Capon	A castrated male chicken.
Carcass	The body of dead animal less the viscera and usually the head, skin and lower legs.

Term	Terminology
Carriers	A edible material to which ingredients are added to facilitate uniform incorporation of the latter into feeds.
Cartilage	The connective tissue attached to the end of bone.
Cast ewe	An aged ewe culled from breeding flock.
Castration	The process of removal of testicles.
Catabolism	The conversion of complex substances into more simple compounds by living cells.
Catalyst	A substance that speed up the rate of a chemical reaction but is not itself used up in the reaction.
Cattle	Members of the cow species-cow, bull, bullock, heifer calves etc.
Chaff	(Part) Glumes, husks, or other seeds covering together with other plants part separated from seed in threshing or processing.
Chemical oxygen demand (COD)	The amount of oxygen required for the oxidation of organic and inorganic matter by a strong oxidizing agent under acidic conditions.
Chevon	Meat of goats.
Chicken	Domestic fowls including chicks, hens, pullets, cockerals, cocks.
Chipped, Chipping process	Cut or broken into fragments; also meaning prepared into small thin slices.
Chooped, Chopping process	Reduced in particle size by cutting with knives or other edged instrument.
Chuni	The compound consists primarily of the broken pieces of endosperm including germ and a portion of husks obtained as by-product during the processing of pulse grains for human consumption.
Closed gilt	A female pig after craniotomy or young pregnant gilt.
Cob	A short age legged, stocky and small horse.
Cock	A mature male chicken.
Cockeral	A male fowl below one year of age.

Term	Terminology
Coefficient of digestibility	The percentage value of a food nutrient that has been absorbed.
Cold slaughter	Carcass from dead of some cause, other than slaughter.
Collagen	Fraction for the fibers of connective tissue, cartilage of bone, forms gelatin when boiled.
Colostrums	Milk of cow for first 3-4 days after calving is known as colostrums. It contains 3 to 4 time more protein, minerals and vitamins also work as laxative.
Colt	A young male horse usually over one year age.
Colt-Foal	An ale foal less than one year age.
Combustion	The combination of substances with oxygen accompanied by the liberation of heat.
Comfort Zone	The temperature range at which no demand is made on temperature regulating mechanism.
Commercial Feed	The term "commercial feed" means all materials which are distributed for use as feed or for mixing in feed.
Compensatory growth	Faster than normal weight gains following a period of under-feeding.
Complete Feed	A nutritionally adequate feed for animals by specific formula is compounded to be fed as the sole ration
Concentrates	They are feeds which contain less than 18% of crude fiber and having more than 60% TDN.
Congenital	Existing at birth.
Congenital Defects	They may be hereditary or non-hereditary but are found at the time of birth, e.g., free martin, cleft palate, white heifer disease, article in animal etc.
Coprophagy	The ingestion of own feaces.
Counter feit	Description of cattle of good colour giving the impression of good breed that they do not possess.
Cow	A female of bovine species that has calved at least once.
Creatnine	A nitrogenous compound arising from protein metabolism and secreted in the urine.
Creep feeding	System of feeding young goats prior to meaning.

Term	Terminology
Crimped, Crimping process	Rolled by use of corrugated rollers. It may entail tempering or conditioning and cooling.
Crips/downers	An animal that has been hurt or crippled.
Critical temperature (lower)	That temperature below which heat production increases in response to a fell in environmental temperature.
Crock	An old ewe.
Crone	An old broken mouthed ewe retained in breeding flock beyond time due to excellent breeding performance. Ewes about 7 to 8 years old when teeth becomes loose.
Crossbreds	The animal produced by mating two different breeds of same species.
Crude fiber	The more fibrous, less digestible fraction of a feed and consists primarily of cellulose, hemicelluloses and lignin.
Crude protein	Total ammonia nitrogen x 6.25, based on the fact that feed protein on the average contains 16.0 per cent nitrogen.
Crumbled, Crumbeling Process	Pellets reduced to granular form.
Crumbles physical form	Pelleted feed reduced to granular form.
Cryptarchidism	Undecided condition of the testes in scrotal sac. The male is called crypt orchid or Rigs.
Cryptarchidism	Undecided condition of the testes in scrotal sac. The male is called crypt orchid or Rigs.
Cub	Young one of a tiger.
Dam	The cow that is served "Female Parent".
Deacon	A young calf not matures enough for veal.
Decortications	Removal of the bark, hull, husk, or shell from a plant seed.
Deficiency Diseases	Nutritional deficiencies lead to some animal diseases, e.g., rickets, night blindness, ostiomalesia etc.

Term	Terminology
Degree Day	65° F minus mean temperature of the day (F).
Dehulled, Dehulling (process)	Having removed the outer covering from grains or other seeds.
Dehydrated, Dehydrating (process)	Having been freed of moisture by thermal means.
Dermatitis	Inflammation of the skin.
Dermatology	Study of integument system (skin)
Desiccate	To dry completely.
Dextrin	An intermediate polysaccharide product obtained during starch hydrolysis.
Diarrhoea	Abnormal frequency and liquidity of fecal discharge.
Diet	Feed ingredient or mixture of ingredients including water, which is consumed by animals.
Digestibility Co-efficient	It is the percentage of total amount consumed, which is digested and absorbed.
Digestible Energy	Represented by that portion of energy consumed which is not excreted in faeces. DE = GE FE. Where, FE = Faeceal Energy.
Digestible energy	The part of the gross energy of a feed which does not appear in the feaces.
Digestion	Processes involved in the conversion of feed into absorbable forms.
Diluents (Physical form)	An edible substance used to mix with and reduce the concentration of nutrients and/or additives in make them more acceptable to animals, safer to use, and more capable of being mixed uniformly in a feed.
Dinmont	Ram lambs after first shearing.
Diverticulitis	Inflammation of a pouch of the intestine.
Doe	An adult female goat, rabbit, or deer.
Doe/Nanny	A female goat which has produced young one at least once.
Doeling	A female kid from the first day of birth until first parturition.

Term	Terminology
Dog	Any member of canine, usually used for male dog. Kitten/Kit: Young one of a cat.
Domestication	Process of adapting of life in intimate association with and to the advantages of man.
Donkey	Synonymous to ass.
Double-rig	A horse of 1 year age or more who's both testes have been retained in the abdomen (Cryptorchidism).
Down-Calve/Springer	A female ox nearly ready to give birth to its calf.
Downeres	Same as crips.
Dry Matter	The part of feed which is not water, sometimes referred to as total solids. This is the sum of the crude protein, crude fat, crude fibers, nitrogen free extract and ash.
Dystokia	Difficulty at the time of parturition
Dystrophy	Degeneration.
Edema	Swelling of a part of or the entire body due to an accumulation of an excess of water.
Educated livestock	Animals consigned from agriculture college/ university.
Elastin	The base of elastic tissue.
Emaciated	An excessively thin condition of body.
Emulsifier (part)	A material capable of causing fat or oils to remain in liquid suspension.
Endemic	Occurring in low incidence but more or less constantly in a given population.
Endocrine	Pertaining to internal secretions.
Endocrinology	Study of Endocrine system (Ductless glands).
Endogenous	Generated within body example, endogenous fecal calcium denotes the calcium in the feaces that is the body and not from the food.
Endometrium	The mucous membrane that lines the uterus.
Energy	The capacity to perform work.
Enteritis	Inflammation of the intestine.

Term	Terminology
Entire	An adult uncastrated horse which is not used for breeding purpose.
Enzyme	One of a class of organic compounds, formed by living cells, capable of producing or accelerating specific organic reaction. An organic catalyst.
Epidemic	When many people in a given region are attacked by some disease at the same time.
Essential Amino Acids	Amino acids as the one of which cannot synthesized in the body of a rate required for normal growth
Esthesilogy	Study of sensory system (Eye, Ear).
Estrous (Heat)	It is a period during which animal strongly, desired matting with a male partner.
Etiology	The causes of a disease of disorder.
Ewe	An adult female sheep which has produced a kid at least once.
Exogenous	Originating from outside of the organism.
Farrowing	An act of delivery in swine or Parturition in pigs.
Fattening cattle	Those nearly ready for butcher.
Fauna	The animal life present, Frequently refer to the overall protozoa population present.
Fecund	Capable of producing many young ones.
Fecundity	It is the potential capacity of the female to produce functional ova, regardless of what happens to them after they are produced.
Feed additive	A material added to animal rations, but which may not supply essential nutrients.
	Is an ingredient or combination of various ingredients added to the basic feed to fulfill the specific need?
Feeding Standards	These are tables' standard amount of various nutrients that should be present in the daily ration of different classes of live stock for optimum result in growth, work production.
Fertility	It is the ability of an animal to produce large number of living young ones.

Term	Terminology
Fill	Coarse feed and water given to animals to increase sale weights.
Filly	A young female horse over one year age.
Filly foal	A female goal below one year age.
Fistula	An opening, often surgically prepared for nutritional studies. For example: rumen fistula or esophageal fistula.
Flakes (Physical form)	An ingredient rolled or cut into flat with prior steam conditioning.
Flavoured Milk	Milk to which some flavours have been added when the milk is used, the product contain milk fat percentage at least equal to the minimum legal requirement for market. But when the fat level is lower (1-2%) the term drink is used.
Flock	A group of goat.
Flora	The plant life present. In nutrition generally refers to the bacteria presents in the digestive tract.
Foal	A young one of either sex up to one year age.
Foaling	An act of delivering a young female horse.
Fold	Enclosure of sheep.
Follicular Phase	Both the protestors and estrous phase of estrous cycle are the part of which known as follicular phase because there is predominance of the hormone estrogen during these two phases.
Food	Edible substance which satisfies the instinct of hunger is food. Food for animal is known as feed.
Free Martin	A heifer calf born as a twin with male most of these heifers is found to be service when they become adult.
	When twin calves of different sexes are born, the female one is usually sterile which is called "Free Martin".
	When female calf is born as twin with male calf if become sterile. Free martin is female in which the reproductive organs have failed to develop properly.

Term	Terminology
Gammen/Wam	The thigh of pig.
Gelatinizing (Process)	A process where starch is ruptured by a combination of moisture, pressure, and in some instances by mechanical shear.
Geld	A castrated male horse of any age.
Gelding (Geld)	A castrated male horse of any age.
Gestation Period	It is the period between the days of conception to the day of calving.
Gestation Ration or Pregnancy allowance	The additional ration for proper growth of fetus after 5th mouth of progeny @ 1.25 and 1.75 kg to Indigenous and crossbred cattle/bushels, respectively.
Gilt	A young female pig not farrowed at least once. She becomes a sow after farrowing.
GiltiMilt/Yeltter	A young female pig kept for breeding purpose which has either not conceived as yet or going to farrow for first time.
Gimmer	A yearling ewe. Or Ewe after first shearing till her first lamb is weaned.
Gluconeogesis	Formation of glucose from protein or fat.
Glycogen	A polysaccharide with the formula which is formed in the liver and muscle and depolymerized to glucose to serve as a ready source of energy when needed by the animal. Also called as animal starch.
Glycogenesis	Conversion of glucose into glycogen.
Glycogenolysis	Conversion of glycogen into glucose.
Glycolysis	The sequence of reaction from glucose to pyruvic acid that is common to carbohydrate catabolism under both anaerobic and aerobic conditions.
Goatlings	A female goat between 1 and 2 years of age.
Goiter	Enlargements of the thyroid gland located in the neck and caused by an iodine deficiency.
Goitrogens	Substances which interfere with thyroxin production in the thyroid gland.
Gonads	The sex-cells are produced by specilised glands known as gonads.

Term	Terminology
Grab Sample	A single sample taken at neither a set time nor a specific flow.
Gras	Acronym for the phrase "generally recognized as safe." A substance, which is generally recognized as safe by experts qualified to evaluate the safety of the substances for its intended.
Grits (part)	Coarsely ground grain, from which the bran and germ have been removed, usually screened to uniform particle size.
Gross Energy	Amount of heat/energy released through the complete combination feed ingredient.
Growing pig	Between for first time.
Hard feeders	Heifers that have lost their figures having had one calf.
Hatching	To bring forth young from the eggs by natural or artificial incubation.
Heat increment	The heat which is unavoidably produced by animal incidental with nutrient digestion and utilization was originally called work of digestion.
Heifer	A female ox of over one year, which has not calved. After calving she becomes a cow.
Heifer calf	Female calf under one year age.
Heiferettes	Dairy type steers that cannot be fattened economically.
Herbivore	An animal that eats primarily plant material.
Herd	Group of animals of cow class.
Hibernation	A dormant state in which some animal pass the winter, in which body temperature drops to slightly above freezing and metabolic activity falls to a very low level.
Hinny	A hybrid whose sire is a stallion dam is a female donkey. It is also sterile.
Hog	A male pig after being castrated, called Barrow.
Hog/Stag	A castrated male pig.

Term	Terminology
Hogg/Hoggests/Tegs	A lamb after weaning but before first shearing (weaning to shearing) or whether lambs when ready for sale.
Hogget	A sheep of either sex between or first shearing. Homogenization The homogenization consists in the sub division of the fat globules into cause the milk to lose its creaming property.
Homeostasis	Maintenance of uniformity and stability
Homoeothermic	Warm-blooded animals in which the body temperature relatively constant.
Hood	A shaped inlet designed to capture contaminated air and conduct it into the exhaust duct system.
Hormone	A chemical substance secreted by an endocrine gland which has a specific effect on the activities of other organs.
Hybrid vigour	A hybrid gains in comparison to either of the parents.
Hybridization	The process of mating animals of two species.
Hydrophobic	Water hating.
Hydrophylic	Water loving.
Hygiene	The science of preserving health by removing or reducing the hazards, associated with the domestic life in more or less artificial environment to enhance animal well-being.
Hypocalcaemia	Decreased level of calcium in the blood.
Hypokalemia	Decreased level of potassium in the blood.
Hen	A mature female chicken.
Immiscible	Not mixing.
Implantation	Embedding of the developing embryos in the living of utters
In situ	Within the body.
Incubation	Hatching of eggs by means of natural or artificial heat.
Ingredient	The term "feed ingredient" means each of the constituent materials making up a commercial feed.

Term	Terminology
In-Iamb	A ewe that is pregnant.
Injuries	Damages done to body due to cuts, bums, blows, falls, accidents etc.
Jack pots	Mixed lot of cattle usually of common quality.
Jejunum	The middle protein of the small intestine which extends from the duodenum to the ileum.
Keratin	A sulphur containing protein which is the primary component of epidermis, hair, wool, hoof, horn, and the organic matrix of the teeth.
Ketosis	A condition characterized by an abnormally elevated of ketone (acetone) bodies in body fluids and tissues-acetonemia.
Kibbled, kibbling (Process)	Cracked or crushed baked dough, or extruded feed that has been cooked prior to or during the extrusion process.
Kid	A young one of goat of either sex exceeding one year.
Kidding	Parturition in goats.
Lamb	Young one of sheep of either sex male lamb known as ram lamb while female lamb known as ewe lamb.
Lamb hog	A male lamb from weaning time until it is shorn
Lactation period	The period between the dates of calving to the date of drying.
Lambing	Parturition in sheep.
Lignin	An indigestible compound which along with cellulose is a major component of the cell wall of certain plant materials such as wood, hulls, straws, and overripe hays.
Limited feeding	Feeding animals to maintain weight and growth but not enough to fatten or increase production. Feeding animals less than they would like to eat.
Limiting amino acid	The essential amino acid of protein that shows the greatest percentage deficit in comparison with the amino acids contained in the same quantity of another protein selected as standard.
Lipids	A broad term for fats and fat-like substance.

Term	Terminology
Litter	A collective term used for all young members of each whelping.
Live stock	Those come in the class of mammalian, *viz.,* cow, buffalow, sheep, goat swine, horse etc. which are domesticated or pet are considered as live stock.
Livestock	Domesticated animals kept on the fann-cattle, buffaloes, sheep, goats, pigs, horses, camels etc.
Maiden ewe	A ewe that has not been mated.
Maiden goat	An unbred goat.
Maiden heifer	An adult female that has not been allowed to breed.
Maintenance Ration	It is the food required by an animal to keep it in good bodily without gain or loss in weight and maintains body temperature.
Maintenance Ration	Amount of feed required to maintain essential body processes of optimum level without gain or loss in body weight and body composition.
Mare	An adult female horse, which produced a young one at least once.
Market milk	It refers to fluid whole milk that is sold to individuals usually for direct consumption. It excludes milk consumed on the farm and that used for the manufacture of dairy organized handling.
Mash (Physical form)	A mixture of ingredients in meal form. Similar term: Mash feed.
Meal	A feed ingredient having a particular size somewhat larger than flour.
Medicated Feed	Any feed which contains drug ingredients intended or represented for the cure, mitigation, treatment, or prevention of disease of animals other than man or which contain drug ingredients intended to effect the structure or any function of the body of animals other than man.
Metabolic body size	An expression relating energy metabolism to body surface. It is usually expressed as a power of body weight such as $W^{0.75}$ or some other exponent.

Term	Terminology
Metabolic Diseases	Sometimes metabolic processes in animal body get disturbed and disease condition, e.g., milk fever, acetonemia/ketosis (due to upset in CHO metabolism).
Metabolic Energy	When energy losses in urine and combustible gases subtracted from digestible the remaining energy is called metabolic energy.
Metabolism	The sum total of the chemical changes in the body including the building up (anabolic, assimilation) and the breaking down process. Transformation by which energy is made available to body use.
Metabolite	Any substance produced by metabolism.
Metabolizable Energy	Digestible energy minus the energy of the urine and fermentation gases.
Micro-Ingredients	Vitamins, minerals, antibiotics, drugs, and other materials normally required and measured in small units.
Milieu	Mixture.
Milk	Milk may be defined as the whole, fresh, clean, lacteal secretion obtained by the complete milking of one or more healthy milch animals, excluding that obtained within 15 days before or 5 days after calving or such periods as may be necessary to render the milk practically colostrums-free, and containing the minimum prescribed percentages of milk fat and milk-solids-not-fat.
Miscellaneous	Diseases Cannot be classified under heads, e.g., Prolapse of vagina bloat, horn cancer etc.
Miscible	Mixable.
Mithun	A semi-nocturnal animal reared at high altitude.
Mucosa	Mucous membrane, mucosal surface of the intestine on the absorptive side.
Mudders	Cattle fed in a muddy feed lot and always penalized in price.
Mule	A hybrid whose size is a donkey and is a mare it is usually sterile.

Term	Terminology
Mule	A hybrid whose sire is a male donkey and dam a mare.
Mutton	Meat of sheep.
Mycology	Study of muscular system (muscles)
Mycotoxin	A fungus or bacteria toxin sometimes present in feed material.
Myoglobin	Muscle hemoglobin.
Myosin	Muscle fiber fraction that complex with actins to bring about muscle contraction.
N.B.	Any cow, a dam and any bull a sire so long as both of them are mated together.
Naism	Dwarf growth.
Nanny	A young female goat.
Net Energy	This is the net reminder of useful energy after all the losses accounted for faeces, urine, and heat increment are subtracted from gross energy.
Net Energy	This is that part of metabolizable energy over the use of which the animal has complete control. It is metabolizeable energy minus heat increment.
Net Protein Ratio	The difference between the average final body weights of a test group of animals fed a protein diet and that of a control group receiving a protein free diet divided by the amount of protein taken by the test group.
Nitrogen Free Extract	The part of feed dry matter, which is not crude protein, crude fat, crude fiber or ash. It consists mainly of sugar and starches. Sometimes referred as NFE.
Non-polar	Immiscible with water.
Neurology	Study of Nervous system (Brain, spinal cord, nervous).
Nutrient	An element or compound that is required in the diet of a given animal to permit normal functioning of the life process.

Term	Terminology
Nutrient	Defined by Morrison as any food constituent or group of food constituents of the same general chemical composition that aids in the support of Animal life.
Omnivore	Literally means eats everything. Animal that eats both plant and animal origin materials.
Open gift	A young female pig yet not served.
Opisthotonus	A condition in which the head tilts back, which may be caused by a deficiency of thiamine. Also called star-gazing.
Ornithology	The study of birds which are not as poultry is known as ornithology.
Ossification	The process of bone formation.
Osteoblasts	Bone forming cells.
Osteoclasts	Bone resobring cells.
Osteology	Study of skeletal system (Bones).
Osteomalacia	A weakening of the bones due to calcium, phosphorous, or vitamin D deficiency.
Osteoporosis	An abnormal porousness of the outermost horny layer of the skin.
Ovine	Animals of subfamily Ovidae.
Ovulation	Occurring the met estrous phase of estrous cycle mature follicle of the ovary ruptures and ovum is released. This process is known as ovulation.
Ovulation	Discharge of egg from the grafian follicle.
Ox	Same as cattle.
Parakeratosis	Any abnormality of the outermost horny layer of the skin.
Parturition	Parturition is the expulsion of the fetus and its membranes from the uterus through the birth canal by natural forces and in such a state of development that foetus is capable of the independent life.
Pedigree bull	The bull whose ancestral record is known.

Term	Terminology
Pig	A domesticated animal *"Susdomesticus"* applicable to every member.
Piggery	Place where pigs are housed.
Pigging/Piggy	Sow that has appearance of having been "suckled" pigs.
Piglets	Small pigs.
Plam shank	The hock end of ham. Pork: Meat from pigs.
Polipeds	Animals having defect hoof.
Pony	A horse of smaller size.
Porker	Pigs whose dead weight ranges between 27 and 36 kg.
Poultry	Domesticated species of birds reared for egg, meat or feathers includes chickens, ducks, turkey, etc.
Prebiotics	Short chain fructose oligosaccharides and mono oligosaccharides encourage growth of beneficial bacteria and inhibit growth of harmful bacteria.
Probiotics	Parker (1974) organism and Substances which contribute to intestinal microbial balance. Fuller (1989) Live microbial feed supplements which beneficially affect the host animal by improving its intestinal microbial balance
Production Ration	It is the food requirement by an animal in addition to the maintenance ration supply the nutrients needed for some forms of production, viz., growth, fattening, milk etc.
Prolificacy	It is used to denote weather many or few off springs result from a given mating or from a certain individual during its lifetime.
Puberty	A stage of physical growth of the body at which the functional gametes are produced for the first time by the animal after its birth is known as puberty. But generally, all related processes necessary for successful reproduction are not started right at puberty.
Pullets	Female birds of 18-26 weeks age.
Puppy or Pup	Young one of a dog species.

Term	Terminology
Queen	A female cat.
Ram	An adult uncastrated male sheep used for breeding purpose.
Ram	A male sheep which is not castrated, Tnp.
Rannies	Small southern calves of poor quality.
Ration	It is often used interchangeable with diet but it may also mean a daily supply of food or feed.
Reproduction	A process of producing or creating a new individual similar to the parents.
Residual Milk	The residual milk is the amount of milk left in the udder after a normal milking. It can obtain only after the injection of oxytocin and remilking the animal.
Rig	A male sheep that has not been properly castrated.
Rig and Rigling	A male horse of above one year age/Cryptorchid whose testes are retained in the abdomen.
Roughages	They are bulky feeds containing relatively large amount of less digestible material, i.e., crude fiber more than 18% and low in the TDN (about 60%).
Runt	Smallest piglet farrowed in the last.
Sand Check	Eggs with a crack in the shell so small that it can be detected only at by candling.
Sausage	Product prepared from fresh minced pork.
Scrub bull	It is a non-descript type of stray village bull.
Seggy	A ram castrated after service.
Semen	Semen is a suspension of spermatozoa in seminal fluid; it is white to light cream coloured fluid.
Semen	Suspension of spermatozoa in seminal fluid.
Seminal Plasma	It is a composite mixture of fluids secreted by organs which in the higher species comprise the epididymis, the vasdeferens, ampullaceal, prostate, seminal vesicles, and cowpers glands and contains other glands located in the wall of the urethral canal.
Service Period	It is the period between the dates of calving to the date of next successive service.

Term	Terminology
Serving	Act of mating/copulation of male and female ox used as the cow has been served by bull. Also used as the service has been given by a bull.
Sexual Maturity	The stage of growth at which the animal can reproduce successfully is called sexual maturity.
Sheep	Used to denote any member of Genus "Ovis".
Shoat/shote	A young pig of either sex between 27 and 72 kg.
Sire	The bull that serves a particular cow (male parent).
Skip	A worthless lamb.
Slanchnology	Study of Digestive, Respiratory, Urinary and Reproductive system.
Slink calves	Calves which have been aborted or those which are found in the uterus during slaughter.
Soliped	Animals having uncleftted hoof-horse.
Sow	An adult female pig used for breeding purpose. Stallion A mature uncastrated house used for breeding purpose.
Springer cow	Cows due to freshen soon.
Stag	A male castrated sheep after it had developed full secondary sexual characteristics.
Stage, Steg or Seg	A male castrated late in life.
Stallion	An uncastrated male used for breeding.
Steer	Male bovine castrated before maturity.
Stickers	Animals difficult to sell.
Stillers	Cattle fed on distillery mash.
Stirk	Applied either to a heifer or a young bullock from about 15 months to 2 years of age.
Stock bull	One used for breeding purpose.
Stock cattle	Young steers and cows that are light in eight and usually immature.
Stocker cattle	Cattle that must be fed out to kill profitably.
Store	Young beast of either sex which are to be fed for butcher later on.

Term	Terminology
Stud bull	Same as breeding bull.
Sty	House of pig.
Swine	All type of domestic pigs.
Teaser ram	A vasectomised ram, sexually active but no service will be fertile.
Term	Explanation
Tnp	An uncastrated adult male sheep.
Tom cat	An uncastrated male cat.
Tonned Milk	It refers to milk obtained by the addition of water and skim milk. It should contain minimum 3% fat
Total Digestible Nutrient (TON)	Per cent digestible protein + % digestible nitrogen free extract + % digestible crude fiber + % digestible other extracts x 2.25.
Toxicity	Poisoning due to eating certain poisonous substances.
Trutters	The feet of sheep, goat/pig.
Tupping	An act of matting male and female sheep.
Vasectomy	It is an act of cutting down the portion of vas-deferens and lighting it raving the blood and nerves supply to testicles infect.
Waler/Whaler	An Australian saddle horse imported into India and other Eastern countries.
Wallow	Water pool for pigs.
Wean	A condition when young calf is no longer fed milk from dam.
Weaner	Piglets separated from mother for tearing independently.
Wedder	A castrated male sheep.
Wet cows	Milking cows.
Wether/Wedder	A castrated adult male sheep.
White faces	Hereford cattle or fine woolen sheep.
Wooden	Cattle, sheep or lambs improperly fed with low grade inferior meat yield.
Yearling	A horse between 1 and 2 years age (yearling colt/filly).
Yearlings	Uncastrated bovine between one and 2 years of age.

IMPORTANT KEY POINTS TO BE REMEMBER ALL THE TIME

1. A 60 gm egg is made up of 10% shell, 30% yolk and 60% white.

2. A cattle disease communicate to man is Anthrax.

3. A chronic buller or the cow having nymphomania will not usually breed until it passes one or two normal heat periods.

4. A condition in cattle called teartness has been identified as molybdenum poisoning.

5. A cradle or bull ring is put in suckling cows.

6. A good layer is among the first off the perch in the morning and among the last on it at night.

7. A group of animals related by descent and similar in most characters like general appearance, features, size, configuration, etc., are said to be a breed.

8. A hormone of adrenal cortex called aldosterone regulates the sodium metabolism by promoting its reabsorption from kidney tubules.

9. A hundred hens will drink 3-6 gallons of water, depending on size of birds, rate of production, salt contents of the ration, and weather conditions.

10. A medium sized egg contains 6.0 gm protein, 6.5 gm fat, 14 vitamins and various minerals.

11. A ram should be use for breeding purpose at 18 months of his age.

12. A sheep is able to bite individual blades of grass unlike a cattle beast because it embodies split upper lip.

13. A sheep require 1.7 sq meter of floor space under a shed.

14. A well built up deep litter in operation for about one year has been found to contain 3, 2 and 2 per cent of nitrogen, phosphorus and potassium, respectively.

15. Addition of aureomycin feed supplement to rations of fattening lambs has been found to reduce the incidence of enterotoxaemia or Over-eating disease.

16. Among the bye-product of grains, rice bran contains the maximum (13.2 per cent) of fat content.

17. Amrit Mahal breed of cattle have compact body frame with short straight back, well arched ribs, sloping quarters.

18. Amrit Mahal breed of cattle have grey coloured body with dark head, neck, hump and quarters. Amrit Mahal breed of cattle have well developed dewlap and hump, very small sheath and close skin.

19. Amrit Mahal is one of the best draught breeds of India. An early moulter under normal conditions is a poor layer.

20. Animal require 60-70 lit water for drinking.

21. Annual laying capacity of an Indian hen is only about 60 eggs which is less than the half of average world is 130 eggs per year.

22. Anthrax is also called spleenic fever.

23. Aseel (poultry) embodies noble qualities of fighting, is a superior sitter and has good parental qualities.

24. At birth, pigs weight between 0.75· 1.75 kilograms.

25. At the junction of omasum and abomasums is an arrangement of folds of mucous membrane, the *vela terminalia,* derived from omasum in the cow, but from the abomasums in the sheep.

26. Australia has the largest sheep population in world.

27. Average milk production of exotic cow is 2000 to 3000 lit/lactation.

28. Back crossing is not widely used by animal breeders.

29. Beginning of early moult signals the lack of persistency.

30. Between 300 and 500 gallons of blood must be passed through the udder for every gallon of milk.

31. Bhadawari breed of buffaloes is found scattered in surroundings of Jamuna and Chambel rivers.

32. Bhadawari buffaloes have medium size and wedge shaped body.

33. Birds grow relatively faster and record higher yield, when kept in a 24 hours lighted house.

34. Bovine veneral trichomoniasis is caused by single celled protozoa, *Trichomones fetus* which brings about the destruction of embryo within 3-5 weeks after conception.

35. Branding of animals should be done at an early age, preferably before the weaning.

36. Broodiness in birds is the external evidence of the maternal instinct and is a dominant sex linked character.

37. Brucellosis (Bang's disease) caused by *Bruella abortus* is usually found in pregnant uterus, but may also localize in udder or testes.

38. Brucellosis is a common genital disease in swine and is dangerous to man.

39. Buffalo milk contain on an average 8.17% fat. Bullocks of Deoni breed are well suited for heavy work.

40. Cannibalism in hens may be by inheritance, deficiency in ration, and by close confinement.

41. Capons are the male birds whose testicles have been removed either surgically of chemically.

42. Castration ameliorates palatability of meat, accelerates the rate of increase in weight and improves the quality of skin is goats.

43. Castration enhances the rate of gain, enables to produce more desirable type of meat and makes the animal more docile.

44. Castration with the help of a Burdizzo's castrator is also known as bloodless castration.

45. Cockerels from 6-8 weeks of age and weighing about 0.5-1.0 kilograms are considered the best subjects for the caponizing operation.

46. Colostrum contains vitamin A.

47. Colostrum is rich in protein (globulin), minerals and antibiotics.

48. Colostrums contains higher proportion of globulin than does normal milk, it contains 3-5 and 5·15 times more protein and vitamin A, respectively.

49. Comparatively small head bulging towards horns, short but stout legs, hooves of hind quarters more backward than those of four quarters, short but well barrel, slightly broad forehead deep in the middle, are main characteristic of Bhadawari breed of buffalo.

50. Cow milk contains on average 3.5 to 5% fat, 13 to 14% total solids and 3.5% protein.

51. Cows can be rebred in 9-12 weeks after parturition. Criss-cross breeding is proposed for utilizing the heterosis in both dam and progeny.

52. Death rates in the inbred animals are higher than those in out bred ones.

53. Deficiency of phosphorous causes pica in cattle.

54. Dehorning of a calf should be done before it is 10 days old because horn button upto this age remains unattached to skull.

55. Diesturm or dioestrum periods are transmitted by activities which normally occur at the end of pregnancy such as nest making and the secretion of milk and these periods are known as Pseudo pregnant periods.

56. Dietary protein may be measured either as crude protein or as COP.

57. Dislike feed, shaking of tail and restlessness etc., are heat signs in doe (she goat).

58. Double-yolked eggs may be due to ripening of two ova at the same time or one ovum being pushed back into the oviduct at the same time when another ovulation takes place.

59. Ears of Deoni (Dongarpatti) breed of cattle have no notch near the tip.

60. Egg shell is mode of ($CaCO_3$).

61. Eighty per cent (80%) incomes of dairy animals come from milk.

62. Estradiol, estrone, estariol, equilin and equienin are called estrogen.

63. Ewes like cows need streaming-up to give good milk yield in early lactation.

64. Ewe's milk on an average has a much higher solid content than cow and goat milk.

65. Exposure of ewes to artificial light brings them to heat, but darkness induces late maturity and reduces oestrus.

66. Fatty sires are often indifferent and may even embody low fertility.

67. Fecundity is the potential capacity of a female to produce function ova, regardless of what happens to them after they are produced.

68. Feed the calf enough of colostrum for the first four days following its birth.

69. First crosses of Sahiwal cow and Ayrshire bull were high milk yielders, but susceptible to F.M. disease.

70. Flushing denotes low ground sheep farming practice of increasing fertility by pulling the ewes on to a higher plane of nutrition before introducing them to the rams.

71. General colours in Sahiwal cattle are various shades of red, pale red and dark brown splashed with white.

72. Grading is the practice of breeding sires of a given breed to non-descript females and their off spring for generation after generation.

73. Guinea grass is a native of Tropical Africa.

74. Gunn in Australia developed a method of semen collection from rams by electric stimulation of spinal cord.

75. Hariana breed of cattle have proportionate body and compact graceful appearance.

76. Hariana cattle breed animals have small and sharp ears, fine textured skin close to the body.

77. Hariana, Ongole, Tharparkar and Kankrej are dual purpose breeds of Indian cattle.

78. Hens with coarse, phlegmatic, masculine, or beefy heads are not likely to lay many eggs.

79. Horse has the largest and most complex large intestine of any of the domestic animal.

80. Hundred Ewes and 3 Rams produced 65 lambs.

81. Hybrids of distant crosses like horse x ass = male and horse x zebra = zebroid are mostly sterile.

82. Hypocalcaecia or milk fever in sheep is often associated with hyomagnesamia.

83. If light provided is more than required it will be lead to cannibalism activity in birds.

84. In cows and she buffalo's cessation of estrus, sluggish temperament, easier tractability and increase in body weight are some of the signs of pregnancy.

85. In dairy cattle the period of ovulation seems to be 14 hours after the end of heat.

86. In guinea-pigs vaginal plug is formed due to semen coagulation after mating.

87. In ruminants, the rumen is the largest, while the reticulum is the most cranial compartment in the stomach.

88. In tail to tail system, of barn arrangement, any sort of minor disease or any change in the hind of the animals can be detected quickly and seen automatically.

89. In the bull the spermatozoon is about 80 mu in length (head is 9 mu long and 5.5 mu wide and about 65 mu in length).

90. In the milk of ruminants, fatty ads are also synthesized in the mammary cells from acetate and about 25 per cent of them are derived from fatty acids coming from dietary fat.

91. Increased deficiency of essential fatty acids markedly enhances the water consumption in animals.

92. Indian goats commonly kid twice in one year and usually thrice in two years.

93. It has been observed that Nili and Ravi breeds of buffalo are distinct without difference and are officially treated as one breed.

94. It has been recommended that sows should be mated on the second of heat for the optimum fertility.

95. It is the face and forehead of Nili and Ravi buffaloes that distinguish this breed mainly from the Murrah.

96. Jaffarabadi breed of buffaloes in their purest form are seen in Gir forest of Kathiawar where they are bred in large numbers, almost solely for *ghee* (butter-oil) production.

97. Kangayam breed of bullocks/bulls are excellent type for hard work.

98. Kangayam breed of cattle are also known as Kongu or Konganad.

99. Kankrej breed of cattle are also known as Banniai or Vagadia or Wadhiar

100. Kankrej cattle have broad chest, their forehead is dished in the centre, have strong curved horns accompanied by skin upto some length, powerful body and straight back.

101. Karakul breed of sheep is famous throughout the world for pelt production.

102. Kellner's idea of feeding animals on the basis of net energy rather than by TDN was first recognized in Germany in 1905 and in England in 1909.

103. Lactose content in different animals' milk is 3 to 5.5% and women milk is 6%.

104. Lambs that are weaned from their mothers and go for further feeding prior to slaughter are known as store lamb.

105. Light in a poultry house does not only destroy disease germs and supplies Vitamin O, but also makes the birds happy.

106. Long sweeping horns of Amrit Mahal breed are typical of Mysore type cattle.

107. Loose horns are common in females of Sahiwal breed of cattle.

108. Lucerne (alfalfa) besides protein is also rich in calcium and phosphorus.

109. Mastitis, in sheep is most commonly caused by *Staphylococcus aureus.*

110. Meat or blood spots may result due to degenerated blood in general or of clots due to haemorrhages of small blood vessel in the ovary or oviduct.

111. Medium and dark brown poultry eggs hatch better than light brown eggs.

112. Milk "let down" mechanism is greatly influenced by sensory and efferent motor sympathetic nerve fibers.

113. Milk is deficient in iron.

114. Milk production enhancement programme is known as Operation Flood.

115. Milk production increase rapidly following parturition, reaches a peak 2-4 weeks where it remains for a short period and then gradually declines.

116. Milk yield and egg production are sex-linked traits. More than 40% of the wool produced in the world comes from the merino and their derivatives.

117. Most of the Indian animal breeds are well known for their draught resisting qualities and to withstand diseases and parasites.

118. Most of the mares are bred to foal in spring, the length of estrus cycle is 15-24 days and the heat period is 3·7 days.

119. Most of the milk is stored in alveoli.

120. Murrah buffaloes have deep massive frame with short, broad back and a comparatively light neck and head. Nellore is the tallest breed of sheep in India.

121. Murrah buffaloes have short, characteristic tightly curled horns, well developed udder and a long tail with a white switch reaching the fetlock.

122. National research center for meat and meat products is situated in Hyderabad (A.P.).

123. Nili and Ravi breeds of cattle are medium sized, have deep frame with an elongated, coarse and heavy head bulging at the top and depressed between the eyes.

124. Normally ovulation occurs about 30-35 minutes after laying each egg.

125. One kg concentrate per 1000 Ibs of body weight is the thumb rule for calculating the required concentrate for cattle.

126. One kg per every 2.5 lits of milk to calculate the concentrate requirement for milch animals.

127. One of the first diluents of semen in veterinary science for the insemination of cattle in Russia was Milovanov's "SGE-2 dilutor".

128. Ongole breed animals are greatly alert and docile with good gait.

129. Ongole breed of cattle is also known as Nellore.

130. Oxytocin and Vasopressin are produced by hypothalamic neurosecretory cells.

131. Parathormone provide optimal level of calcium, while insulin regulates carbohydrate metabolism during lactation period.

132. Parturition is presumably caused by an increase in the level of estrogen, a decrease in the level of progesterone, increase in the sizes of fetus, and inadequacy of nutrients.

133. Pashmina is hairy under coat obtained from Pushmina goat.

134. Penicillin is more effective than other antibiotics as feed supplement to poultry especially for growing chicks and turkey poults.

135. Peninsular region has the highest sheep population in India followed by North Western region their wool production is 28 and 64 per cent respectively.

136. Perosis disease of chickens is due to the deficiency of manganese.

137. Phillips and Lardy (1939-40) introduced an advance technique of semen storage known as egg yolk phosphate diluents.

138. Pituitary has frequently been described as the conductor of the endocrine orchestra.

139. Poultry egg contains 12.4% protein, 10.7% fat.

140. Project Directorate on cattle is situated in Modipuram (U.P.).

141. Prolificacy is more or less restricted in its application to the female or groups such as breed, strain or heard. It Ram produce more wool than the ewes but quality of wool is inferior.

142. Reversion of vagina its protrusion out through the vulva is known as ballooned vagina.

143. Rinderpest is also called as Cattle Plague.

144. River buffaloes originate from the Indian subcontinent, have chromosome number of 2n = 50 and are improved domesticated animals with high milk and meat production.

145. Sahiwal and Sindhi are muilch breeds of Indian cattle.

146. Sahiwal is also called as Lola.

147. Sahiwal is one of the best dairy breeds in India.

148. Semen is a suspension of spermatozoa in seminal fluid; it is opaque, white to light cream-coloured fluid.

149. Seventy eight to eighty three (78 to 83%) per cent of egg laid before noon and 1-2% after 3 p. m.

150. Severe riboflavin deficiency in chicks causes swelling and softening of the sciatic and brachial nerves producing a contraction of toes and typical symptoms known as curled-toe paralysis.

151. Shantagurtadia breed is well known for meat purpose.

152. Sharp horns, relatively small ears, well placed hump covered with a tuft of hair at the top, strong legs and feet, and tight sheath are some of the characters of Siri breed of animals.

153. Sheep is a golden hoofed animal.

154. Sheep is seasonally poly-gastrous animals.

155. Sheep require 3 to 4 kg of dry matter per 100 kg of their body weight.

156. Shell membrane is formed in the Isthmus.

157. Sindhi breed of cattle has medium size and is compact; it has well proportioned body and is extremely docile.

158. Station (male horse) x Janned (female ass) = Hinny which is generally inferior to mule.Tharparkar breed of cattle are also known as Thari.

159. Stiff lamb disease caused by deficiency of vitamin E.

160. Tharparkar breed of cattle are medium sized, deep built and embody short, straight and strong limb.

161. The active principle of the thyroid gland was first isolated by Kendall who named it thyroxin, found in a protein material called thymoglobulin.

162. The age of bull at maturity will be 3 years of village condition.

163. The animal specially cattle diseases *viz.,* "salt sick" and "lechsucht" are caused by copper deficiency.

164. The anterior pituitary secretes F.S.H. in increasing quantities bringing about the multiplication of puberty.

165. The antibiotic vitamin B_{12} feed supplement is largely included in the commercial formula feed for chicks, pigs and dairy claves for their growth and fattening.

166. The average annual milk production per cows and buffalo are 157 kg and 504 kg respectively.

167. The average percentage increase in homozygosity or decrease in hetrozygosity is an inbred animal in relation to an average animal of the same breed of the foundation stock is known as the co-efficient of inbreeding.

168. The average specific gravity of whole semen is 1.028, 1.011 and 1.03 in man, dog and bull, respectively.

169. The best known and commonly used method of semen collection is Artificial Vagina (AV) method.

170. The best known breeds of Indian buffaloes are Murrah, Jaffarabadi, Nili and Bhadawarl.

171. The best time for cutting a crop for hay making is when it is 30-50% in blossom.

172. The boiling point of milk is 100.17°C.

173. The carbohydrate of sugar constituent of the milk is lactose.

174. The cells which supply nutrition to the maturing spermatide are known as sertoli's cells.

175. The colostrum contains greater percentage of protein, a large part of which is due to the presence of gamma globulins the antibody proteins.

176. The colour of the Kangayam bull is of grey with grey to black markings, while the cow is white with black marking just in front of the fetlocks on all the four legs sometimes on the knees.

177. The deficiency of F.S.H. and L.H. hormone may limit or retard reproduction in either sex.

178. The deficiency of fat leads to a number of troubles in animals namely, lack of libedo in males, re-absorption of the fetus, abortion or dead birth etc.

179. The Digestibility co-efficient may be defined as the percentage of the total amount consumed which is digested and absorbed.

180. The ears of Gir breed of cattle are markedly long, pendulous resembling a tiny curled leaf.

181. The eggs having a proportion of white to yolk of about 2:1 usually hatch better than those having either wider or narrower ratio.

182. The embryonic mortality may be due to hormone deficiency or imbalance, failure of implantation of fertilized ova, death due to lethal genes, accidents in development, insufficient nutrition, overcrowding or infection in the uterus.

183. The female organ extending from the cervix posteriorly upto the urogential sinus or vestibule from which it is separated by the hymen or the hymenal constriction is the organ of copulation in females.

184. The first breed of poultry named Plymouth rock was exhibited at America's first poultry show, held at Boston in 1849.

185. The general colour of the Malvi breed of cattle is grey to iron grey black in the neck and quarters.

186. The Gir (also called Kathiawarhi Sort or Desan) breed of Indian cattle probably originated in the Gir forests of South Kathiawar.

187. The Government of India started artificial insemination centers at Calcutta, Hissar, Madras, Bangalore and Nagpur in 1945.

188. The hens kept 10 battery system produce vegetative eggs.

189. The hens start laying eggs at about 15 days age. At this age 14 hours light is necessary.

190. The home of Sindhi breed of Indian cattle is Karachi and Hyderabad (Sindh) in Pakistan.

191. The hormone called testosterone has been considered to be the primary sex hormone.

192. The Indian Veterinary Research Institute, Izatnagar, started experimental work on artificial insemination under the direction of Dr. P. Bhattacharya, in 1942.

193. The Kangayam breed of animals Rave moderate sized hump, wide muzzle strong limbs, small dewlap, fire skin, very small sheath, well developed quarters and a fine tail.

194. The Kankrej breed of cattle is one of the heaviest of Indian breeds.

195. The length of estrous cycle for heifers is 20 days and 21 to 22 days for cows.

196. The main problem in horse breeding is the usually long heat period and the variation in ovulation time.

197. The Malvi breed bullocks are good for road and field work, economical feeders and have good adaptability.

198. The Malvi breed of cattle have deep, short and compact body, their ears are alert and short in size, horn short but thick tapering to a blunt point.

199. The Malvi breed of cattle is also known as Mahadespuri or Manthai.

200. The mammary secretion for the first three or four days following calving is called beestings.

201. The maternal dystokia may be due to uterine abnormalities like torsion of uterus, uterine intertia, and uterine hernia.

202. The mixture of grasses in a good pasture should have high percentage of leguminous grasses (forage crops).

203. The most popular fine-wool breed in the world is Merino.

204. The most reliable method of pregnancy diagnosis in bovine animals is palpation of the uterus per rectum.

205. The Murrah breed of buffalo is the most efficient producer of milk, not only in India but probably in the world.

206. The noticeable feature of Jaffarabadi buffalo-breed is the very prominent forehead and heavy horns, inclined to droop on each side of the neck and then turn up at the points.

207. The organisms *Fusarium nodosus* attacks only sheep and goats.

208. The ovaries of a sow are rounded and the surface is lobulated with follicles and corpora lutea.

209. The oviduct (fallopian tube or ovarian tube) has a diameter of about 0.1 cm in case of the cow and it widens into a funnel shaped tube at the ovarian end which is called the as infundibulum.

210. The ovulation in cows takes place by the action of luteinizing hormone.

211.   The percentage of thick white in egg tends to decrease from about April to July in most of the flocks.

212.   The popular colour of murrah buffaloes is jet black with white markings on the tail, face and extremities.

213.   The portion termed as germ spot on the yolk of the egg gives rise to a blastoderm which ultimately develops into an embryo.

214.   The poultry breed Rhode Island Red is one of the hardiest of all the breeds.

215.   The purity of the breed is maintained by confining the mating of animals to within the breed.

216.   The ram is in full vigour for breeding during his age of 2.5 to 5 years.

217.   The recommended calcium phosphorus ratio for reproduction in pigs is approximately 1.5:1.

218.   The retractor muscle in bulls which is attached to the anterior arms of the flexure helps in pulling the erected and exposed penis back into the sheath for protection against injury and infection.

219.   The silent or shy breeder cows and buffaloes go through the normal ovarian changes in the estrous cycle but show no signs of heat and receptivity to males.

220.   The sire is half of the heard.

221.   The Siri breed cattle have massive body, small head, square cut, wide and flat forehead, presenting no convexity.

222.   The Siri breed of cattle is found in Darjeeling hill tracts, Sikkim and Bhutan.

223.   The spermatozoa are generated in the testes and stored in epididymis, where seminal plasma is contributed by the secretary fluids produced in accessory organs.

224.   The sperms, in case of a hen, remain in the oviduct for 2-3 weeks after mating, but the newest sperm is the one most likely to fertilize the egg.

225.   The stages of parturition (a continuous act) comprises the stages, viz., preliminary dilation fetus expulsion and after-birth stage.

226.   The suckling stimulus and intra-mammary pressure are two important factors influencing the level at which lactation is maintained.

227.   The swam water buffalo *(Bubolus bubalis)* is indigenous to southeast Asia, southeast China and Assam in India.

228.   The systematic cross breeding of sheep on a regional scale is called stratification.

229. The tail of Tharparkar breed of cattle is fine with black witch.

230. The term class is used to designate groups of poultry breeds developed in certain areas.

231. The term concentrate is applied to that group of feeds which are relatively high in total digestible nutrients and low in crude fiber.

232. The thick walled portion of the reproductive tract lying between the uterus and the vagina is known as cervix.

233. The udder secretion, immediately before parturition changes from honey like secretion to a yellowish one, known as colostrum.

234. The unborn calf makes most of its growth during the last 2 to 3 months before birth.

235. The yellow colour of egg yolk is due to the pigment xanthophylls.

236. Twenty five of the total world cattle are presented in India and milk production is only 6%.

237. Twenty five point thirty (25·30) sq. ft. area is considered sufficient and economical for housing dairy cattle.

238. Two medium sized eggs supply about 26% of iron 20% of phosphorus, 10% of iodine, 8% of calcium and 20% of protein of the daily requirement of an adult male.

239. Two types of pigments in milk are carotene and riboflavin.

240. Two-toothed females (sheep) are referred to as gimmers or shearling eves, while the males as shearling rams.

241. When a female calf is born as a twin to a male calf, a sterile female is known as Freemaretin or neuter.

242. White leghorn is the most popular breed for egg purpose in the world originated in Italy.

243. Wool of Patanwadi is suitable for manufacturing of carpet and blankets.

244. Yellow colour in milk is due to carotene and yellowish green due to riboflavin.

245. Zinc deficiency in cattle causes parakeratosis.

4

SHORT EXPLANATIONS

Importance of Livestock in Agriculture

India is basically a rural-oriented and land-based (76.27%) rural population. Being an agricultural country with 1/5th of the world's population of cattle and half of the world's buffaloes, the cows and bullocks are the backbone of the agriculture and play a major role in the rural economy. Most of the farmers are poor with small holdings of land which is cultivated with the help of bullocks on which they depend for ploughing, planking, irrigation, manuring, threshing, transport of produce etc. Cows provide milk and milk products which are the only sources of animal protein in vegetarian diets. The transport of agricultural products from villages to the market is done by bullocks. Poor economic condition of the farmer does not permit him to use artificial fertilizers and tractors for cultivation. The country has an obligation to the cows, their off springs and work of bullocks. Plants grown as fodder must supply all the necessary nutrition for animals. The substance which the animals require from their feeds is known as nutrients, namely, proteins, carbohydrates, fats, minerals, vitamins and water, for several purposes (to maintain her own weight, growth, production of milk, reproduction and work etc.).

The investment in animal husbandry and dairy development is less than 10 per cent of the investment in the agricultural sector, but its contribution to the GNP from agriculture and animal husbandry is reported to be more than 31 per cent (Acharya, 1989). This includes the contribution for draught animal power.

Milk production in India is characterized by low yield non-descript cows and buffaloes, millions of small producers with little or no holdings, use of crop residues and natural herbage with or without costly concentrate as feed supplement and scarce land for pastures and forage production. In areas with balanced mixed farming, milk production is usually higher. The best animals are found where agriculture has been prosperous and cultivated fodder and cereals and oilseed milling byproducts in addition to crop residue are readily available.

The advantages of livestock in India are as follows:

1. Milk and milk products from dairy animals are palatable and easy to digest, therefore an important human food.

2. Milk contains fat, proteins, milk, sugar, minerals and vitamins in a much better proportion and quantity than any other food; so it is nearly perfect food.

3. Milk being the only source of animal protein in vegetarian diet; it is a protective and balanced food.

4. Livestock supplies meat worth Rs. 1200 million per year.

5. Source of draught power as bullocks is used for various agricultural operations.

6. Due to small-sized holding livestock fits well in mixed farming.

7. Provides organic manure for maintenance of soil fertility.

8. Offers opportunity of earning foreign exchange worth Rs. 800 million annually.

9. Contributes 7% to national income.

10. Makes better use of agricultural, industrial, dairy and animal by-products.

11. Milk output accounts for 5.6% GNP.

12. Milk output accounts for more than 17% of India's agricultural production.

Making Dairying More Profitable

Dairying cannot become profitable unless determined attention is paid to milk and calf production. The dairy farmers suffer losses due to various causes such as deaths of animal, infestation with worms, outdated husbandry practices, self-medication of sick animals, unvaccinated stock and poor breeding. A healthy livestock is fundamental to the welfare of a nation as it provides milk, meat, hide, draught power and fuel. It provides stability to agricultural-industrial economy. The entire edifice of dairying rests on four pillars-feeding, breeding, heeding and weeding.

Feeding : The dairy must take it for granted that one of the most important parts of dairying is sound feeding the animal. Cross breeding for higher productivity coupled with more and more understanding of feeding and management has put greater pressure on dairy animals. Besides producing large quantities of milk, a dairy cow is supposed to carry her next calf because "a calf a year" is an essential action plan, for higher production and profit. The dairy owner has a great choice for locally available feed and food, which need to be supplemented with minerals, vitamins and trace elements.

Well-balanced ration plays an important role in improving the performance and health of the dairy animal. It must be remembered that overfeeding is as harmful as underfeeding. Much progress is being made in the field of animal nutrition. The dairy farmer, besides making the best use of his experience and observation, should remain in close contact with the animal nutritionist who is in a better position to convey the latest on the scientific feeding. The sound feeding of the dairy animal is not a simple matter. The feed and fodder should be fibrous, scientifically balanced, economical and palatable.

Breeding : The regular breeding and genital disease-free herd is cherished by a prospective dairy farmer. The tool to improve livestock quality and production depends on artificial insemination (AI) of local cows and buffaloes with the semen of the bulls of high genetic potential. The AI as a means of milk and calf production is now accepted and utilized worldwide. By this method, several thousands of females are inseminated artificially with semen collected from bulls and maintained at semen collection centers. Since 5,000 to 10,000 doses of semen can· be processed from a single bull, it becomes imperative to ensure that bulls donating semen with impaired fertilizing capacity are not used. For this, regular evaluation of each and every bull is necessary. The evaluation requires a gynecologist to conduct a physical examination, a bacteriologist to conduct disease tests and a semenologist to evaluate semen. Unless the infertility in males' arid females is dealt properly, the AI system of breeding will not yield the desired results. All dairy farmers should avail the expertise of the state bacteriologist to get their whole herd examined for Brucellosis and other allied genital diseases known to cause infertility and abortions in cows. When semen is of good quality and female to be inseminated is free from genital defects, the proficiency of inseminator matters. His ability, experience and technique play a vital role in achieving cherished conception rate.

Heeding : Nothing is more unfortunate than the occurrence of an outbreak of a contagious disease in the herd of a dairy farmer. Besides spending a huge sum of money on buying medicines for the treatment of sick animals, the farmer also has to suffer the loss of milk. Sometimes, he has to bear the brunt of the deaths of costly animals. It should be a routine with dairy farmers to vaccinate their animals against contagious diseases well in advance. The prophylactic vaccines for most of the contagious diseases are freely available in the market. These vaccines are extra-fragile and as such due care is needed in procurement and vaccination. The animals to be vaccinated should be free from worms for the optimum production of antibodies against the diseases. Protection against contagious diseases and parasites (external and internal) will ensure health and efficiency of the dairy animals.

New health technologies can play an effective role in the treatment of various diseases of animals. At the village level, inadequate veterinary service is available to dairy farmers. Many vets, especially the less qualified, are known for pricking unnecessary injections. This is due to the ineffective diagnostic facilities and consequent treatment by hit-and-trial methods. This is unfair in the modern scientifically advanced era. Veterinary medicine has gone through substantial changes during the last century. Major technological advances are still not available to dairy farmers. This is the reason for low profile treatment of sick and costly dairy animals. The ultimate sufferer is the dairy owner.

Weeding : Timely disposal of animals suffering from incurable diseases like tuberculosis should be done to save time, labour and money spent on their management and feeding.

"A healthy livestock is fundamental to the welfare of a nation as it provides milk, meat, hide, draught power and fuel. It provides a stable agricultural industrial economy".

Plants as feed for Cow

Milk contains the same compounds as cow's feed, however in different forms. Dairy cow produces her milk from the feed and water what she consumes. Unless she gets all these as enough nutrition she cannot be expected to produce large amount of milk. Plants to grow well as feed for animals, require the elements from soil, which if deficient will not be able to supply enough and make up right kind of feed. Therefore, to maintain its fertility regular manuring or fertilization is necessary. For improving the water-holding capacity, good texture and stable fertility the manure obtained from animals is added.

Animal husbandry in India

Animal husbandry includes domestication of animals to obtain animal products like milk, meat, wool, skin and Hyde etc. and to use them for draught and transportation. These animals are cow, buffalo, goat, sheep, pig, camel, horse, mule, donkey and yak etc. India has about 500 species of animals of which only few are domesticated for different uses.

Cattle

India has about 20 per cent of the world's cattle population. These animals are the backbone of the country's agriculture and have significant con-tribution in rural economy. Bullocks have major role in agricultural operations and rural goods movement and transportation while cows provide nutritious milk to enrich Indian diet. These are also good source of hides and skins for leather industry which earns substantial foreign exchange. Also cow dung is a good source of manure and domestic fuel. According to 1997 live stock census there were 198.9 million cattle in the country of which 42 per cent were bullocks, 32 per cent cows and 26 percent young livestock. There has been 28.1 per cent increase in the number of cattle between 1951 and 1997. At state-level Bihar has the largest percentage (12.37) of cattle in the country followed by Uttar Pradesh (10.06), Madhya Pradesh (9.80) Maharashtra (9.09) and West Bengal (8.97). These five states together provide about 50 per cent of the country's cattle number. Sikkim (0.07%), Arunachal Pradesh (0.23), Nagaland (0.19), Meghalaya (0.38), Manipur (0.26), and Tripura (0.62) have less than 1 per cent of the cattle in the country.

The density of cattle population per 100 hec-tares of total cropped area in India is about 105 which vary from 295 in Jammu and Kashmir to 32 cattle in Punjab. Here six states of the country (Manipur, Jammu and Kashmir, Megh Bihar. Himachal Pradesh and Tripura are characterized by the highest density (over 200 cattle/100 of gross cropped area) of cattle. Similarly states (Haryana. Punjab, Rajasthan Gujarat) record the lowest cattle density (less 80 cattle/100 ha of gross cropped area). Remain' states fall under the medium cattle dense category. Cattle population in India belongs to differ: breeds. These include: (i) milch breed, (ii) draw' breed, and (iii) mixed or general breed.

1. Millch Breeds

Here cows yield higher quantity of milk but the bullocks are not of good quality. Some important milch breeds include Gir, Sindhi, Sahiwal, Tharparkar and Deoni. The Gir breed is a native of Saurashtra ' which yields about 3175 kg of milk per lactation period. Sahiwal breed belongs to Montgomery dis-trict of Pakistan yielding 2725-4535 kg of milk per lactation period. The Sindhi and Red Sindhi breeds hail from the Sindh area of Pakistan producing about 5440 kg of milk per lactation period. The Deoni breed belongs to the western and north-western parts of Andhra Pradesh where cow yields 1580 kg of milk; per lactation period. The Tharparkar breed is also a native of Sindh area of Pakistan whose cow yields 1815 to 2720 kg of milk per lactation period.

2. Draught Breeds

Here the cows are poor milkers but the bul-locks are excellent draught animals. This group consists of (a) short-horned, white or light grey colour with coffin-shaped skull and face slightly convex in profile, e.g. Nagori and Bachaur. (b) the lyre horned grey coloured with wide forehead, promi-nent orbital arches, flat or dished profile, deep body and powerful draught capacity, e.g., Kathiawar, Malvi and Kherigarh; (c) the Mysore type characterized by prominent forehead with long and pointed horns which rise close together, e.g. Mallikar, Amritmahal, Kangyam and Killari; and (d) small black, red or dun coloured with large patches of white markings, found in the foot hill region of the Himalayas, e.g., Ponwar and Siri.

3. Dual Purpose Breeds

Here cows are fairly good yielders of milk and the bullocks are good for draught purposes. The group includes: (a) short-horned, white or light grey cattle with long coffin-shaped skull and face slightly convex in profile, *e.g.,* Mariana, Ongale, Gaolo, Rath, Dangi, Krishna Valley and Nimari etc; and (b) lyre-horned, grey cattle, deep bodied with wide forehead, prominent arches, flat or dished in profile and good draught capacity, *e.g.,* Tharparkar and Kankrej. The Mariana breed is very popular in Haryana, Delhi and western Uttar Pradesh.

Bullocks are strong and useful for draught purposes and cows yield up to 5 kg. of milk per day. The Ongale belongs to Guntur and Nellore districts of Andhra Pradesh whose bullocks are heavy ploughing and carting. The Gaolo breed is indigenous to Nagpur and Wardha districts of Maharashtra and Chhindwara district of Madhya Pradesh whose cows yield about 7.5 kg of milk every day. Rath breed is an admixture of the Mariana, Nagori and Mewati breeds. Its cows give up to 5 kg of milk per day and the bullock is fit for draught work. The Dangi breed comes from Nashik, Thane, Ahmadnagar and Kolaba districts of Maharashtra. The Krishna Valley is very popular breed of north Karnataka and southern Maharashtra. Its cows provide about 916 kg of milk per lactation period while bullocks are good for agricultural work. The Nimari breed is very common in East and West Nimar districts of Madhya Pradesh whose cows yield about 915 kg of milk per lactation period. The Kankrej breed is indigenous to the Gujarat plains whose cows provide 4.5 to 6.5 kg of milk per day and the bullocks are sturdy for draught work.

In order to improve the breed of the Indian cattle 7 central cattle breeding farms have been established. Some of the exotic breeds yielding higher quantity of milk like Jersey, Holstein-Friesian, Swiss-Brown, Gurnsey, German Fleckvich and Ayreshire have been introduced in the country which is becoming popular amongst dairy farmers. The maximum lactation milk yield of these cross breeds has been 6,000 kg While the average milk-yield is about 2,600 kg.

DIFFERENT TYPES OF CATTLE BREEDS

List of Cattle Breeds

The following is a list of breeds of cattle. Over 800 *breeds* of *cattle* are recognized worldwide, some of which *adapted* to the local *climate*, others which were bred by humans for specialized uses. Breeds fall into two main types, regarded as either two closely related *species*, or two *subspecies* of one species. *Bos indicus* (or *Bos taurus indicus*) cattle, also called zebu, are adapted to hot climates. *Bos taurus* (or *Bos taurus taurus*) are the typical cattle of Europe, north-eastern Asia, and parts of Africa – they are referred to in this list as "taurine" cattle, and many are adapted to cooler climates. *Taurus/indicus* hybrids are widely bred in many warmer regions, combining characteristics of both the ancestral types (such as the *Sanga cattle* of Africa). In some parts of the world further species of cattle are found (both as wild and domesticated animals), and some of these are related so closely to taurine and indicus cattle that inter*species hybrids* have been bred. Examples include the *Dwarf Lulu* cattle of the mountains of *Nepal* with *yak* blood, the *Beefalo* of North America with *bison* genes, the *Selembu* breed of *India* and *Bhutan* with *gayal*genes. The *Madura* breed of *Indonesia* may have *banteng* in its parentage. The *Dzo* of Nepal is an infertile cattle-yak crossing which is bred for agricultural work. Like the *mule* which is also infertile, they have to be continually bred from the parent species.

Breed	Subspecies	Country/region of origin	Meat	Dairy	Draught	Other
Abigar	Taurus	-	-	Dairy	-	-
Abondance (cattle)	Taurus	France	Meat	Dairy	-	-
Abyssinian Shorthorned Zebu	Indicus	Ethiopia	-	-	Draught	-
Aceh	-	Indonesia	-	-	-	-
Achham	Indicus	Nepal	-	Dairy	-	-
Adamawa (cattle)	Taurus	Nigeria	Meat	Dairy	Draught	-
Adaptaur	Taurus/Indicus hybrid	Australia	Meat	-	-	-
Afghan	-	-	-	-	-	-
Africangus	Taurus/Indicus hybrid	United States	Meat	-	-	-
Afrikaner	Taurus/Indicus hybrid	South Africa	Meat	Dairy	-	-
Agerolese	Taurus	Italy	-	Dairy	-	-
Akaushi	-	Japan	Meat	-	-	one of the four Kobe breeds
Ala Tau	Taurus	Kazakhstan	Meat	Dairy	-	-
Alambadi	-	India	-	-	-	-
Albanian (cattle)	Taurus	Albania	-	Dairy	Draught	-
Albanian Dwarf	-	-	-	-	-	-
Alberes	Taurus	-	Meat	-	-	-
Alderney	Taurus	Channel Islands	-	Dairy	-	-
Alentejana	-	-	-	-	-	-
Alentejana Cattle	Taurus	Portugal	Meat	-	Draught	-
Aleutian wild cattle	-	United States	-	-	-	Feral
Aliad Dinka						
-	-	-	-	-	-	
Alistana-Sanabresa	-	-	-	-	-	-
Allmogekor	Taurus	Sweden	Meat	-	Draught	-
Alur	-	-	-	-	-	-
American	Taurus/Indicus hybrid	United States	Meat	-	-	-
American Angus	-		-	-	-	-
American Beef Friesian	-		-	-	-	-
American Brown Swiss	-		-	-	-	-
American Milking Devon	Taurus	United States	Meat	Dairy	Draught	-

Contd...

Breed	Subspecies	Country/region of origin	Meat	Dairy	Draught	Other
American White Park	Taurus	United States	Meat	Dairy	-	-
Amerifax	Taurus	United States	Meat	Dairy	-	-
Amrit Mahal	Indicus	India	-	-	-	-
Amsterdam Island cattle	Taurus	Amsterdam Island	-	-	-	Feral
Anatolian Black	Taurus	Turkey and Bulgaria	-	Dairy	Draught	-
Andalusian Black	-	Spain	-	-	-	-
Andalusian Blond	-	Spain	-	-	-	-
Andalusian Grey	-	Spain	-	-	-	-
Angeln cattle	Taurus	Germany	-	Dairy	-	-
Angoni	-	-	-	-	-	-
Angus cattle	Taurus	Scotland	Meat	Dairy	-	-
Ankina	-	-	-	-	-	-
Ankole-Watusi	Indicus	Northern Africa	Meat	Dairy	Draught	Show
Aosta cattle	Taurus	-	-	-	Draught	-
Apulian Podolian	-	-	-	-	-	-
Aracena	-	-	-	-	-	-
Arado	-	-	-	-	-	-
Argentine Crillo	-	-	-	-	-	-
Argentine Criollo	Taurus	Argentina	Meat	Dairy	-	-
Argentine Friesian	-	-	-	-	-	-
Armorican (cattle)	Taurus	France	Meat	Dairy	-	-
Arouquesa Cattle	Taurus	Portugal	Meat	Dairy	-	-
Arsi	-	-	-	-	-	-
Asturian Mountain	Taurus	Spain	Meat	Dairy	-	-
Asturian Valley	Taurus	Spain	Meat	Dairy	-	-
Aubrac cattle	Taurus	France	Meat	Dairy	-	-
Aulie-Ata	Taurus	Kazakhstan	-	Dairy	-	-
Aulie-Atta	Taurus	-	Meat	Dairy	-	-
Aure et Saint-Girons	Taurus	France	Meat	Dairy	Draught	-
Australian Braford	Taur hybrid us/Indicus	Australia	Meat	-	-	-
Australian Brangus	Taurus/Indicus hybrid	Australia	Meat	-	-	-
Australian Charbray	Taurus/Indicus hybrid	Australia	Meat	-	-	-
Australian Friesian Sahiwal	Taurus/Indicus hybrid	Australia	-	Dairy	-	-
Australian Milking Zebu	Taurus/Indicus hybrid	Australia	-	Dairy	-	-

Contd...

Breed	Subspecies	Country/region of origin	Meat	Dairy	Draught	Other
Australian Shorthorn	-	-	-	-	-	-
Austrian Simmental	-	-	-	-	-	-
Austrian Yellow	-	-	-	-	-	-
Avetonou	-	-	-	-	-	-
Avileña	Taurus	-	Meat	-	Draught	-
Avilena-Black Iberian	-	-	-	-	-	-
Aweil Dinka	-	-	-	-	-	-
Ayrshire	Taurus	Scotland	-	Dairy	-	-
Azaouak cattle	-	-	-	-	-	-
Azebuado	-	-	-	-	-	-
Azerbaijan Zebu	-	-	-	-	-	-
Azores	-	-	-	-	-	-
Baherie cattle	Indicus	Eritrea	Meat	Dairy	-	-
Bakosi cattle	Taurus	Cameroon	Meat	-	-	Rituals
Balancer cattle	Taurus	-	Meat	Dairy	-	Gelbvieh/ Angus cro-ssbreed
Baoule	Taurus	Ivory Coast	-	-	-	-
Barrosã Cattle	Taurus	Portugal	Meat	-	Draught	-
Barzona	Taurus/Indicus hybrid	United States	Meat	-	-	-
Bazadais	Taurus	France	Meat	-	-	-
Beef Freisian	Taurus	-	Meat	Dairy	-	-
Beefalo	Taurus/Bison hybrid	United States	Meat	-	-	-
Beefmaker	Taurus/Indicus hybrid	-	Meat	-	-	-
Beefmaster	Taurus/Indicus hybrid	United States	Meat	-	-	-
Belgian Blue	Taurus	Belgium	Meat	Dairy	-	-
Belgian Red	Taurus	Belgium	Meat	Dairy	Draught	-
Belgian Red Pied	Taurus	Belgium	Meat	Dairy	-	-
Belgian White -and-red	Taurus	Belgium	Meat	Dairy	-	-
Belmont Red	Taurus/Indicus hybrid	Australia	Meat	-	-	-
Belted Galloway	Taurus	Scotland	Meat	-	-	-
Bernese	-	-	-	-	-	-
Berrenda cattle	Taurus	-	Meat	Dairy	Draught	Sport
Betizu	Taurus	Spain/France	Meat	-	Draught	-

Contd...

Breed	Subspecies	Country/region of origin	Meat	Dairy	Draught	Other
Bianca Val Padana	Taurus	-	Meat	Dairy	-	-
Blaarkop	Taurus	Netherlands	-	Dairy	-	-
Black Angus	Taurus	Scotland	Meat	Dairy	-	-
Black Baldy	Taurus	-	Meat	-	-	-
Black Hereford	Taurus	England	Meat	-	-	-
Black Pied Dairy Cattle	Taurus	Germany	-	Dairy	-	-
Blanca Cacereña	Taurus	-	Meat	Dairy	Draught	-
Blanco Orejinegro BON	Taurus	Colombia	Meat	Dairy	Draught	-
Blonde d'Aquitaine	Taurus	France	Meat	Dairy	Draught	-
Blue Albion	Taurus	Great Britain	Meat	-	-	-
Blue Grey	Taurus	Great Britain	Meat	-	-	-
Bohus Polled	-	-	-	-	-	-
Bonsmara	Taurus/Indicus hybrid	South Africa	Meat	-	-	-
Boran	Indicus	eastern Africa	Meat	-	-	-
Boškarin	Taurus	Croatia, Slovenia	Meat	Dairy	Draught	-
Braford	Taurus/Indicus hybrid	-	Meat	-	-	-
Brahman (cattle)	Indicus	India	Meat	Dairy	Draught	-
Brahmousin	Taurus/Indicus hybrid	-	Meat	-	Draught	-
Brangus	Taurus/Indicus hybrid	United States	Meat	-	-	-
Braunvieh	Taurus	Switzerland	Meat	Dairy	-	-
Brava	-	-	-	-	-	-
British White	Taurus	Great Britain	Meat	Dairy	-	-
British Friesian	Taurus	Great Britain	Meat	Dairy	-	-
Brown Carpathian	Taurus	-	-	-	-	
Brown Swiss	Taurus	Switzerland	-	Dairy	-	-
Bue Lingo	Taurus	United States	Meat	Dairy	-	-
Burlina	Taurus	Italy	-	Dairy	-	-
Buša cattle	Taurus	former Yugoslavia (Dinaric Alps)	Meat	Dairy	Draught	-
Butana and Kenana	Indicus	-	-	Dairy	-	-
Cachena Cattle	Taurus	Portugal/Spain	Meat	Dairy	Draught	-
Caldelana	Taurus	-	Meat	Dairy	Draught	-
Camargue	Taurus	France	Meat	-	Draught	Sport
Campbell Island Cattle	Taurus	New Zealand	-	-	-	Feral

Contd...

Breed	Subspecies	Country/region of origin	Meat	Dairy	Draught	Other
Canadian Speckle Park	Taurus	Canada	Meat	-	-	-
Canadienne	Taurus	Canada	Meat	Dairy	-	-
Canaria	Taurus	-	Meat	-	Draught	-
Canchim	Taurus/Indicus hybrid	Brazil	Meat	-	-	-
Caracu	Taurus	Brazil	Meat	Dairy	Draught	-
Cárdena andaluza	Taurus	-	Meat	-	Draught	-
Carinthian Blondvieh	Taurus	Austria	Meat	Dairy	Draught	-
Carora	Taurus	Venezuelan	Meat	Dairy	Draught	-
Chinese Central Plains Yellow	Taurus/Indicus hybrid	-	-	-	-	-
Charbray	Taurus/Indicus hybrid	-	Meat	-	-	-
Charolais	Taurus	France	Meat	-	Draught	-
Chateaubriand	-	-	-	-	-	-
Chiangus	Taurus	-	Meat	-	-	-
Chianina	Taurus	Italy	Meat	-	Draught	-
Chillingham Cattle	Unknown	England	-	-	-	Feral
Chinese black pied	Taurus	China	-	Dairy	-	-
Coloursided White Back	Taurus	-	Meat	Dairy	-	-
Commercial	-	-	-	-	-	-
Corriente cattle	Taurus	Spain	Meat	Dairy	Draught	Sport
Costeño con Cuernos	Taurus/Indicus hybrid	Colombia	Meat	-	Draught	-
Crioulo Lageano	-	-	-	-	-	-
Dajal (cattle)	Indicus	-	Meat	Dairy	Draught	-
Danish Black-Pied	-	-	-	-	-	-
Danish Jersey	Taurus	Denmark	-	Dairy	-	-
Danish Red	Taurus	Denmark	Meat	Dairy	-	-
Deep Red Cattle	Taurus	-	Meat	Dairy	-	-
Devon	Taurus	England	Meat	Dairy	-	-
Dexter	Taurus	Ireland	Meat	Dairy	-	-
Dhanni (cattle)	Indicus	-	Meat	Dairy	Draught	-
Doayo cattle	Taurus	Cameroon	-	-	-	-
Doela	-	-	-	-	-	-
Dølafe	Taurus	Norway	Meat	Dairy	-	-
Droughtmaster	Taurus/Indicus hybrid	Australia	Meat	-	-	-

Contd...

Breed	Subspecies	Country/region of origin	Meat	Dairy	Draught	Other
Dulong'	Taurus	-	Meat	-	Draught	-
Dutch Belted	Taurus	Netherlands	Meat	Dairy	-	-
Dutch Friesian	Taurus	Netherlands	Meat	Dairy	-	-
Dwarf Lulu	Taurus/Indicus/Yak hybrid	-	-	-	-	-
East Anatolian Red	Taurus	-	-	Dairy	-	-
Eastern Finncattle	Taurus	Finland	Meat	Dairy	-	-
Eastern Red Polled	-	-	-	-	-	-
Enderby Island Cattle	Unknown	New Zealand	-	-	-	Feral
English Longhorn	Taurus	England	Meat	Dairy	-	-
Ennstal Mountain Pied Cattle	Taurus	Austria	Meat	-	Draught	-
Estonian Holstein	-	-	-	-	-	-
Estonian Native	-	-	-	-	-	-
Estonian Red cattle	Taurus	-	Meat	Dairy	-	-
Évolène Cattle	Taurus	Switzerland	-	Dairy	-	-
Finnish Ayrshire	-	Finland	-	-	-	-
Finnish cattle	Taurus	Finland	Meat	Dairy	-	-
Finnish Holstein-Friesian	-	-	-	-	-	-
Fjäll	Taurus	Sweden	Meat	Dairy	-	-
Fleckvieh	Taurus	Austria	-	Dairy	Draught	-
Florida Cracker	Taurus	United States	Meat	-	-	-
French Simmental	Taurus	France	Meat	Dairy	-	-
Fribourg black and white	-	-	-	-	-	-
Friesian Red and White	Taurus	-	Meat	Dairy	-	-
Fulani Sudanese	Indicus	-	Meat	Dairy	Draught	-
Galician Blond	Taurus	Spain	Meat	Dairy	-	-
Galloway cattle	Taurus	Scotland	Meat	-	Draught	-
Garvonesa Cattle	Taurus	Portugal	Meat	-	Draught	-
Gascon cattle	Taurus	France	Meat	-	Draught	-
Gelbray	Taurus/Indicus hybrid	-	Meat	-	-	-
Gelbvieh	Taurus	Germany	Meat	Dairy	Draught	-
Georgian mountain cattle	Taurus	Georgia	Meat	Dairy	Draught	-

Contd...

Breed	Subspecies	Country/region of origin	Meat	Dairy	Draught	Other
German Angus Cattle	Taurus	Germany	Meat	-	-	-
German Black Pied Cattle	Taurus	Germany	-	Dairy	-	-
German Red Pied	Taurus	Germany	Meat	Dairy	-	-
Gir	Indicus	India	Meat	Dairy	-	-
Glan cattle	Taurus	Germany	Meat	Dairy	Draught	-
Gloucester	Taurus	England	Meat	Dairy	Draught	-
Gobra	Indicus	-	Meat	-	Draught	-
Greek Shorthorn	Taurus	Greece	Meat	Dairy	-	-
Greek Steppe cattle	Taurus	-	Meat	-	-	-
Greyman Cattle	Taurus/Indicus hybrid	Australia	Meat	-	-	-
Groningen	Taurus	Netherlands	-	Dairy	-	-
Groningen White-Headed	Taurus	Netherlands	Meat	Dairy	-	-
Gudali	Indicus	-	-	Dairy	Draught	-
Guernsey	Taurus	Channel Islands	-	Dairy	-	-
Guzerat	Indicus	India	Meat	-	Draught	-
Halikar	Indicus	-	Meat	Dairy	Draught	-
Hanwoo	Taurus	Korea	Meat	-	Draught	Cow fighting
Hariana (cattle)	Indicus	-	Meat	Dairy	Draught	-
Hartón del Valle	Taurus	Colombia	Meat	Dairy	Draught	-
Harz Red mountain cattle	Taurus	Germany	Meat	Dairy	Draught	-
Hays Converter	Taurus	Canada	Meat	-	-	-
Heck Cattle	Taurus	-	-	-	-	Science
Hereford	Taurus	England	Meat	-	-	-
Herens	Taurus	Switzerland	Meat	-	-	Cow fighting
Hhybridmaster	Taurus/Indicus hybrid	-	Meat	Dairy	-	-
Highland Cattle	Taurus	Scotland	Meat	-	-	-
Hinterwald Cattle	Taurus	Germany	Meat	Dairy	-	-
Holando-Argentino	Taurus	Argentina	Meat	Dairy	-	-
Red Holstein	Taurus	-	-	Dairy	-	-
Holstein	Taurus	-	-	Dairy	-	-
Horro	Indicus	-	Meat	-	Draught	-
Hungarian Grey	Taurus	Hungary	Meat	-	Draught	-
Iberian cattle	-	-	-	-	-	-

Contd...

Breed	Subspecies	Country/region of origin	Meat	Dairy	Draught	Other
Icelandic	Taurus	Iceland	-	Dairy	-	-
Illawarra cattle	Taurus	-	Meat	Dairy	-	-
Improved Red and White	Taurus	-	Meat	-	-	-
Indo-Brazilian	Indicus	-	Meat	-	-	-
Irish Moiled	Taurus	Ireland	Meat	Dairy	-	-
Israeli Holstein	Taurus	-	-	Dairy	-	-
Israeli Red	Taurus/Indicus hybrid	-	Meat	Dairy	-	-
Istoben cattle	Taurus	-	Meat	Dairy	-	-
Istrian cattle		*see Boškarin*				
Jamaica Black	Taurus/Indicus hybrid	-	Meat	Dairy	-	-
Jamaica Hope	Taurus/Indicus hybrid	Jamaica	-	Dairy	-	-
Jamaica Red	Taurus/Indicus hybrid	-	Meat	-	-	-
Jarmelista Cattle	Taurus	Portugal	Meat	-	Draught	-
Jersey	Taurus	Channel Islands	-	Dairy	-	-
Jutland cattle	Taurus	Denmark	Meat	Dairy	-	-
Kalmyk cattle	Taurus	-	Meat	Dairy	-	-
Kangayam Cow	Indicus	India	Meat	-	Draught	-
Kankrej	Indicus	-	Meat	-	Draught	-
Karan Swiss	Taurus/Indicus hybrid	-	Meat	Dairy	-	-
Kazakh White-headed cattle	Taurus	-	Meat	Dairy	-	-
Kerry cattle	Taurus	Ireland	Meat	Dairy	-	-
Kholomogory	Taurus	Russia	Meat	Dairy	-	-
Kostroma Cattle	Taurus	-	Meat	Dairy	-	-
Krishna Valley cattle	Indicus	-	Meat	Dairy	Draught	-
Kuri	Taurus	-	Meat	Dairy	Draught	-
Kurgan	Taurus	-	Meat	Dairy	-	-
Kuri	Taurus	-	Meat	Dairy	Draught	-
Lampurger	-	-	-	-	-	-
Latvian Blue	-	-	-	-	-	-
Latvian Brown	Taurus	Latvia	Meat	Dairy	-	-
Lebedyn (cattle)	Taurus	Ukraine	-	&mdash		
Levantina	Taurus	-	Meat	-	Draught	-
Limia Cattle	Taurus	Spain	Meat	Dairy	Draught	-
Limousin	Taurus	France	Meat	-	Draught	-

Contd...

Breed	Subspecies	Country/region of origin	Meat	Dairy	Draught	Other
Limpurger	Taurus	-	Meat	Dairy	-	-
Lincoln Red	Taurus	England	Meat	-	-	-
Lineback	Taurus	-	-	Dairy	-	-
Lithuanian Black-and-White	-	-	-	-	-	-
Lithuanian Light Grey	-	-	-	-	-	-
Lithuanian Red	-	Lithuania	Meat	-	-	-
Lithuanian White-backed	-	-	-	-	-	-
Lohani cattle	Indicus	-	Meat	-	Draught	-
Lourdais	Taurus	France	Meat	Dairy	Draught	-
Australian Lowline	Taurus	Australia	Meat	-	-	-
Luing	Taurus	Scotland	Meat	-	-	-
Madagascar Zebu	Indicus	-	Meat	Dairy	Draught	-
Madura cattle	Javanicus/Indicus hybrid	Indonesia	Meat	Dairy	Draught	Racing
Maine Anjou	Taurus	France	Meat	Dairy	Draught	-
Mandalong Special	Taurus/Indicus hybrid	-	Meat	-	-	-
Mantequera Leonesa	Taurus	-	Meat	Dairy	Draught	-
Maramure? Brown	-	Romania	-	-	-	-
Marchigiana	Taurus	Italy	Meat	-	Draught	-
Maremmana	Taurus	Italy	Meat	Dairy	Draught	-
Marinhoa Cattle	Taurus	Portugal	Meat	Dairy	Draught	-
Marinhoa	-	-	-	-	-	-
Maronesa	Taurus	Portugal	Meat	Dairy	Draught	-
Masai cattle	Indicus	-	Meat	-	Draught	-
Mashona	Taurus	-	Meat	-	Draught	-
Menorquina	Taurus	-	Meat	-	Draught	-
Mertolenga Cattle	Taurus	Portugal	Meat	-	Draught	-
Meuse-Rhine-Issel	Taurus	Netherlands	Meat	Dairy	-	-
Milking Shorthorn	Taurus	Great Britain	Meat	Dairy	-	-
Minhota Cattle	Taurus	Portugal	Meat	Dairy	Draught	-
Minhota	-	-	-	-	-	-
Miniature	-	-	-	-	-	-
Mirandesa cattle	Taurus	Portugal	Meat	-	Draught	-
Moc?ni?? (cattle)	-	Romania	-	-	-	-
Monchina (cattle)	Taurus	Spain	Meat	-	Draught	-

Contd...

Breed	Subspecies	Country/region of origin	Meat	Dairy	Draught	Other
Mongolian cattle	Taurus	China	Meat	-	Draught	-
Montbeliard Cattle	Taurus	France	Meat	Dairy	-	-
Morucha	Taurus	-	Meat	-	Draught	Fighting
Murboden Cattle	Taurus	Austria	Meat	Dairy	Draught	-
Murnau-Werdenfels Cattle	Taurus	Germany	-	Dairy	-	-
Murray Grey	Taurus	Australia	Meat	-	-	-
N'Dama	Taurus	Guinea	Meat	Dairy	-	-
Negra Andaluza	Taurus	-	Meat	-	Draught	-
Nelore cattle	Indicus	-	Meat	-	Draught	-
Nguni	Taurus/Indicus hybrid	-	Meat	Dairy	-	-
Normande Cattle	Taurus	France	Meat	Dairy	-	-
Northern Finncattle	Taurus	Finland	Meat	Dairy	-	-
Northern Shorthorn	Taurus	Great Britain	-	Dairy	-	-
Chinese Northern Yellow	-	-	-	-	-	-
Norwegian Red	Taurus	Norway	Meat	Dairy	-	-
Ongole Cattle	-	India	Meat	-	Draught	
Pajuna	Taurus	Spain	Meat	-	Draught	-
Palmera	Taurus	-	Meat	-	Draught	Sport
Pantaneiro cattle	-	Brazil	-	-	-	-
Parda Alpina	Taurus	-	Meat	Dairy	-	-
Parthenais	Taurus	France	Meat	-	Draught	-
Pasiega	Taurus	Spain	-	Dairy	-	-
Pembroke cattle	Taurus	Wales	Meat	Dairy	-	-
Philippine Native cattle	Taurus	-	-	Dairy	Draught	-
Pie Rouge des Plaines	Taurus	France	Meat	Dairy	-	-
Piedmontese	Taurus	-	Meat	Dairy	-	-
Pineywoods	Taurus	-	Meat	Dairy	Draught	-
Pinzgauer	Taurus	Austria	Meat	Dairy	-	-
Pirenaica	Taurus	-	Meat	-	Draught	-
Podolica	Taurus	Italy	Meat	Dairy	Draught	-
Polish Black-and-White	-	-	-	-	-	-
Polish Red cattle	Taurus	Poland	Meat	Dairy	Draught	-
Polled Hereford	Taurus	England	Meat	-	-	-
Polled Shorthorn	Taurus	England	Meat	-	-	-

Contd...

Breed	Subspecies	Country/region of origin	Meat	Dairy	Draught	Other
Ponwar (cattle)	Indicus	-	Meat	-	-	-
Preta Cattle	Taurus	Portugal	Meat	-	Draught	-
Punganur Cow	Indicus	-	Meat	Dairy	Draught	-
Pustertal Pied Cattle	Taurus	-	Meat	-	-	-
Qinchaun	Taurus	-	Meat	-	Draught	-
Queensland Miniature Boran	Taurus	-	Meat	-	-	Pets
Ramo Grande	Taurus	Portugal	Meat	Dairy	Draught	-
Randall	Taurus	United States	Meat	Dairy	Draught	-
Rath (cattle)	Indicus	India[3]	Meat	Dairy	Draught	-
Rathi (cattle)	Indicus	-	Meat	Dairy	Draught	-
Rätische Grauvieh	Taurus	-	Meat	Dairy	Draught	-
Red Angus	Taurus	Scotland	Meat	Dairy	-	-
Red Brangus	Taurus/Indicus hybrid	-	Meat	-	-	-
Red Fulani	Taurus/Indicus hybrid	-	Meat	-	-	-
Red Poll	Taurus	England	Meat	Dairy	-	-
Red Polled Østland	Taurus	-	Meat	Dairy	-	-
Red Sindhi	Indicus	-	Meat	Dairy	-	-
Reina	Taurus	-	Meat	Dairy	-	-
Retinta	Taurus	-	Meat	-	Draught	-
Riggit Galloway	Taurus	-	Meat	-	-	-
Ringamåla Cattle	Taurus	-	-	Dairy	-	-
Rohjan	Indicus	-	Meat	-	Draught	-
Romagnola cattle	Taurus	Italy	Meat	-	Draught	-
Romanian B?l?ata	-	Romania	-	-	-	-
Romanian Steppe Gray	-	Romania	-	-	-	-
Romosinuano	Taurus	Colombia	Meat	-	-	-
Russian Black Pied	Taurus	Russia	Meat	Dairy	Draught	-
RX3	Taurus	-	Meat	-	-	-
Sahiwal	Taurus	-	Meat	Dairy	-	-
Salers	Taurus	-	Meat	-	Draught	-
Salorn	Taurus	-	Meat	-	-	-
Sanga	Taurus/Indicus hybrid	-	Meat	-	-	-
Sanhe	Taurus/Mongolian hybrid	-	Meat	Dairy	Draught	-

Contd..

Breed	Subspecies	Country/region of origin	Meat	Dairy	Draught	Other
Santa Cruz cattle	Taurus/Indicus hybrid	-	Meat	-	-	-
Santa Gertrudis cattle	Taurus/Indicus hybrid	-	Meat	-	-	-
Sayaguesa	Taurus	Spain	Meat	-	Draught	-
Schwyz	-	-	-	-	-	-
Selembu	Gaurus/Indicus hybrid	-	Meat	Dairy	-	-
Senepol	Taurus/Indicus hybrid	-	Meat	-	-	-
Sheko	-	Ethiopia	-	-	-	-
Shetland cattle	Taurus	Scotland	Meat	-	Draught	-
Shorthorn	Taurus	England	Meat	Dairy	-	-
Siboney	Taurus/Indicus hybrid	-	Meat	-	Draught	-
Sided Trönder	-	-	-	-	-	-
Simbrah	Taurus/Indicus hybrid	-	Meat	-	-	-
Simford	Taurus	-	Meat	-	-	-
Simmental	Taurus	Switzerland	Meat	Dairy	Draught	-
South Devon	Taurus	England	Meat	Dairy	-	-
Batangas	Taurus/Indicus hybrid	-	-	-	Draught	-
Spanish Fighting Bull	Taurus	Spain	Meat	-	-	Sport
Speckle Park cattle	Taurus	-	Meat	-	-	-
Square Meater	Taurus	-	Meat	-	-	-
Sussex cattle	Taurus	England	Meat	-	-	-
Swedish Friesian	Taurus	-	-	Dairy	-	-
Swedish Mountain	-	-	-	-	-	-
Swedish Red Cattle	Taurus	-	-	Dairy	-	-
Swedish Red Poll	Taurus	-	Meat	Dairy	-	-
Swedish Red-and-White	Taurus	-	Meat	Dairy	-	-
Symons Type	Taurus/Indicus hybrid	-	Meat	-	-	-
Tabapuã (cattle)	Indicus	Brazil	Meat	Dairy	Draught	-
Tarentaise	Taurus	France	Meat	Dairy	-	-
Tasmanian Grey	Taurus	-	Meat	-	-	-

Contd...

Breed	Subspecies	Country/region of origin	Meat	Dairy	Draught	Other
Tauros	Taurus	-	-	-	-	grazing projects, rewilding
Telemark cattle	Taurus	Norway	Meat	Dairy	-	-
Texas Longhorn	Taurus	United States	Meat	Dairy	-	-
Texon	Taurus	-	Meat	-	-	-
Tharparkar	Indicus	India[4]	-	Dairy	Draught	-
Tswana cattle	Taurus	Botswana	Meat	Dairy	-	-
Tudanca	Taurus	Spain	Meat	Dairy	Draught	-
Tuli	Taurus	Zimbabwe	Meat	Dairy	-	-
Tulim	Taurus/Indicus hybrid	-	Meat	-	-	-
Turkish Grey Steppe cattle	Taurus	-	Meat	-	Draught	-
Tux Cattle	Taurus	-	-	Dairy	-	-
Tyrolese Grey Cattle	Taurus	Austria	Meat	Dairy	-	-
Ushuaia Wild Cattle	-	-	-	-	-	Feral
Ukrainian Grey cattle	Taurus	Ukraine	-	-	-	
Vaca Toposa or Vaquilla	Taurus	-	-	-	-	Sport
Väne	-	-	-	-	-	-
Vaynol cattle	Taurus	-	Meat	-	-	-
Vechoor cow	Indicus	-	-	Dairy	-	-
Vestland Fjord	Taurus	-	Meat	Dairy	-	-
Vianesa	Indicus	-	Meat	Dairy	-	-
Volyn meat cattle	-	Ukraine	Meat	-	-	-
Vorderwald Cattle	Taurus	Germany	Meat	Dairy	-	-
Vosges	Taurus	-	Meat	Dairy	-	-
Wagy?	Taurus	Japan	Meat	-	Draught	-
Wangus	Taurus/Indicus hybrid	-	Meat	-	-	-
Welsh Black	Taurus	Wales	Meat	-	-	-
Western Finncattle	Taurus	Finland	Meat	Dairy	-	-
Western Fjord	-	-	-	-	-	-
Western Red Polled	Taurus	-	Meat	Dairy	-	-
White Cáceres	Taurus	-	Meat	-	Draught	-

Contd.

Breed	Subspecies	Country/region of origin	Meat	Dairy	Draught	Other
White Fulani	Indicus	-	Meat	-	-	-
White Park	Taurus	Great Britain	Meat	Dairy	-	-
Whitebred Shorthorn	Taurus	Great Britain	Meat	-	-	-
Xingjiang Brown	Taurus/Mongolian hybrid	-	Meat	Dairy	Draught	-
Yakutian cattle	Taurus	Russia	Meat	Dairy	Draught	-
Yanbian cattle	Taurus	China	Meat	-	Draught	-
Yurino (cattle)	Taurus	Ukraine	&mdash	-	-	
Zubro?	Taurus/wisent hybrid	-	-	-	-	Science

Breeding category, breeds and cultivars

Breeding Category	Breeds and Cultivars
Methods	• Selective breeding; Crossbreed; Inbreeding; Smart breeding; Mutation breeding; Preservation breeding; • Outcrossing Purebred Hybrid
Animal breeds	• Cat; Cattle; Chicken; Dog • Breeding • Duck; Goat; Goose; Guinea pig; Horse • Breeding • Pig; Pigeon • breeding • Rabbit • Sheep • Turkey • Water buffalo Backyard breeder Breed standard Breed type Breeding back Breeding pair Breeding program Captive breeding Designer crossbreed

Contd...

Breeding Category	Breeds and Cultivars
Cultivars (Lists)	
	Apple, Banana, Basil, *Callistemon*, *Canna*, Cherimoya, Citrus, Coffee, *Gazania*, Grape, *Grevillea*, Hop, Mango, *Nemesia*, *Nepenthes*, *Olives*, *Pear*, Pumpkin, Rice, Rose (breeders, cultivars), Strawberry, Sweet potato, Sweetcorn, Tomato, Plant breeding, Nepal, Tree breeding, Heirloom plant etc.
Selection Methods & Genetics	• Marker-assisted selection, Natural selection, (Balancing, Directional, Disruptive, Negative, Stabilizing,
	• Selective sweep), Culling.
	• **Selection Methods in Plant Breeding**
	Genotype
	Phenotype
	Dominance
	Codominance
	Epistasis
	Dwarfing
	Heterosis
	Outbreeding depression
	Inbreeding depression
	Recessive trait
	Sex linkage
	F1 hybrid
Other	• Breeder; Rare Breed; Landrace; Breed registry

Indigenous Cattle Breeds

Indigenous breeds are as follows

Indigenous Breeds are classified under three groups based on utility / purpose.

(*a*)   Milch breeds / Milk breeds

(*b*)   Dual Purpose breeds

(*c*)   Draught breeds

Milch Breeds / Milk Breeds

The cows of these breeds are high milk yields and the male animals are slow or poor work animals. The examples of Indian milch breeds are shahiwal, Red Sindhi, Gir and Deoni The milk production of milk breeds is on the average more than 1600 kg per lactation.

Dual Purpose Breeds

The cows in these breeds are average milk yielder and male animals are very useful for work. Their milk production per lactation is 500 kg to 150 kg. The example of this group is Ongole, Hariana, Kankrej, Tharparker, Krishna valley, Rathi and Goalo Mewathi.

Draught Breeds

The male animals are good for work and Cows are poor milk yielder are their milk yield as an average is less than 500 kg per lactation. They are usually white in color. A pair of bullocks can haul 1000 kg. Net with an iron typed cart on a good road at walking speed of 5 to 7 km per hour and cover a distance of 30 - 40 km per day. Twice as much weight can be pulled on pneumatic rubber tube carts. The examples of this group are Kangayam, Umblacherry, Amritmahal, and Hallikar.

Cart Pulling Bull   Ploughing Bull

Exotic breed – Milch – Jersey, Holstein Friesian

Milch Breed *Red Sindhi*

- Hailing from the Kohistan, Sindh province in present Pakistan, this breed is one of the most distinctive cattle breeds of india.

- Mainly available in Punjab, Haryana, Karnataka, Tamil Nadu, Kerala and Orissa.

- Under good management conditions the Red Sindhi averages over 1700 kg of milk after suckling their calves but under optimum conditions there have been milk yields of over 3400 kg per lactation.

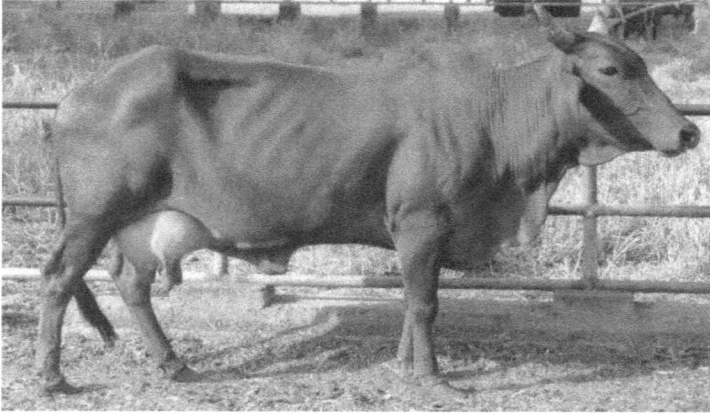

Red Sindhi

Sahiwal

- Originally Belonging to the Montgomery district of Present Pakistan
- Mainly found in Punjab, Haryana, U.P, Delhi, Bihar and M.P.
- Milk yield – Under village condition :1350 kg
- Milk yield – Under commercial farms: 2100 kg
- Age at first calving -32-36 months
- Calving interval – 15 month

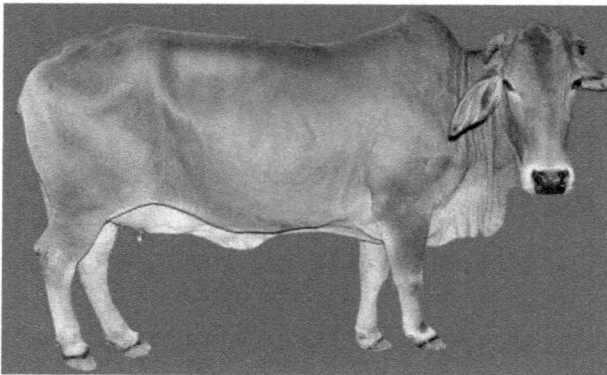

Sahiwal

Gir

- Mainly found in Gir forest areas of South Kathiawar
- Gir Cows are good Milk – yielder

- Milk yield – Under village condition : 900 kg
- Milk yield   – Under commercial farms: 1600 kg

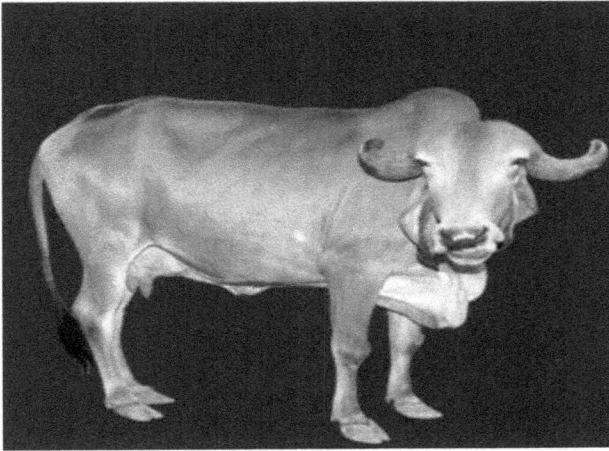

Gir

Deoni

- Mainly found in North western and western parts of A.P.
- Cows are good milk producers and bullocks are good for work

Milch and Draught breeds

Hariana

- Mainly found in Karnal, Hisar and Gurgaon district of Haryana, Delhi and Western M.P Milk yield –1140 -4500 kgs
- Bullocks are powerful for road transport and rapid ploughing

Hariana

Tharparkar

- Mainly found in Jodhpur, Kutch and Jaisalmer
- Milk yield – Under village condition :1660 kg
- Milk yield – Under commercial farms: 2500 kg

Tharparkar

Kankrej

- Mainly found in Gujarat
- Milk yield – Under village condition : 1300 kg
- Milk yield– Under commercial farms : 3600 kg
- Age at first calving -36 to 42 months
- Calving interval – 15 to 16 months
- Bullocks are fast, active and strong. Good for plough and cart purpose

Kankrej

Draught Breeds

Kangayam

- This breed, in its native area, is also known by other names of Kanganad and Kongu though the name Kangayam is well-known. These cattle are bred in the southern and southeastern area of the Erode district of Tamilnadu in India.

Kangayam

- Mainly found in Coimbatore, Erode, Namakkal, Karur and Dindigul districts of Tamil Nadu.
- Best suited for ploughing and transport. Withstands hardy conditions.

Amritmahal

- Mainly found in Karnataka.
- Best suitable for ploughing and transport.

Amritmahal

Hallikar

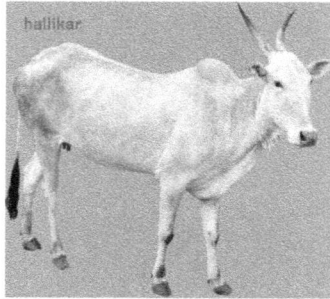

Hallikar

- Mainly found in Tumkur, Hassan and Mysore districts of Karnataka
- Bullocks are strong, well spirited, quick and steady in the field as well as on road.

Umblacherry

Origin: Tanjore district in Tamilnadu.

Distinguishing characters

- This breed has similar characters as kangayam.
- Bulls are fearly temperament. They are used for ploughing in Thanjore delta area.
- Calves are red in colour when born and become grey in colour after 6 months of age.
- Cows are poor milker with average milk yield of 300 kg/lactation.
- Male animals are good for hard work.

Umblacherry

Khillari Cattle

The Khillari is breed of *cattle*, of the *Bos indicus* sub-species, found in Man and Khatav taluka in *Satara* and Shirol taluka in *Kolhapur* and *Atpadi* taluka in *Sangli* in the *Mumbai* region of western *India*. The breed is well adapted to the tropical and drought prone conditions present in this part of the world and are favoured by the local farming community due to their ability to handle the hardships of farming pretty well. In spite of this, lately the breed is showing a steady decline in numbers mostly due the low milk yield which forms an alternate stream of income for the farming community.

The Khillari breed, with its several varieties, possibly owes its origin to the Hillikar breed of cattle from *Mysore* State or from the *Maharashtra* state. The name comes from "Khillar" meaning a herd of cattle, and Khillari meaning the herdsman.Mostly Khillari bulls are basically from *Satara District* of South Maharashtra.& also this animals are found in neighbouring districts of *Sangli,Kolhapur* & *Solapur* of Western Maharashtra.

The Khillari is between 4½ to 5½ feet tall and weighs between 800 and 1000 lbs.The typical Khillari animal is compact and tight skinned, with clean cut features and squarely developed hindquarters. The appearance is compact with stout strong limbs. The pelvis is slightly higher than the shoulders. The Khillaris of the Deccan plateau, the Mhaswad and the Atpadi Mahal types are greyish white in colour, the males having deeper colour over the forequarters and hindquarters, with peculiar grey and white mottling on the face. The Tapti Khillari is white with reddish nose and hooves. The Nakali Khillari is grey with tawny or brickdust color over the forequarters. Newly born calves have rust red coloured polls, but this disappears within a couple of months. Khillaris have a long narrow head with long horns sweeping back and then upward in a distinctive bow, and tapering to a fine point. The ears, coloured yellow inside are small, pointed and held sideways. The legs are round and straight with black hooves. The coat is fine, short and glossy.

There are four principal types of Khillaris prevalent in the different regions of *Maharashtra* State. The variety Hanam Khillar, or sometimes known as *Atpadi* Mahal, is prevalent in southern Maharasthra. In the districts of *Kolhapur* and *Satara* and the adjoining areas the variety known as *Mhaswad* Khillari is mostly in *Man & Khatav* talukas of *Satara District* prevalent. In the area of the Satpura range of hills comprising the West Khandesh district the variety prevalent is known as Tapi Khillari or Thillari. A variety of more recent origin known as Nakali Khillari - Nakali means "imitation" - is found in adjacent areas of these regions.

In the southern Maharashtra and the districts of Sholapur, Sangli and Satara the Khillaris are bred by cultivators. In these regions the size of the herd is small, usually not more than one or two cows. In the Satpura ranges the Khillaris are bred by professional breeders known as Thillaris. These breeders produce bulls and bullocks for which there is always a very good demand. Besides their extensive use in their home tracts they are used in the adjacent districts of Pune. Ahmednager, Nasik and Bijapur. Khillaris are classified as "medium fast draft". Breeding is carried out by the Government of Mahararashtra at Hingoli, Jath and Junoni and by the Government of *Karnataka* at Bankapur.

Khillar

Different Exotic Breeds?

Jersey

Origin: This breed was developed from the island of jersey in the English channel off the coast of France.

Distinguishing Characters

- The Jersey is one of the oldest dairy breeds, having been reported by authorities as being purebred for nearly six centuries
- The color in Jerseys may vary from a very light gray or mouse color to a very dark fawn or a shade that is almost black. Both the bulls and females are
- Commonly darker about the hips and about the head and shoulders than on the body.
- Age at first calving : 26-30 months
- Inter calving – 13-14 months

- Milk yield – 5000-8000 kg
- Dairy milk yield is found to be 20 liter whereas cross bred jersey, cow gives 8-10 liter per day.
- In India this breed has acclimatized well especially in the hot and humid areas.

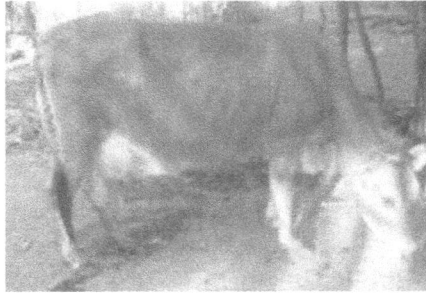

Jersey

Holstein Friesian

Origin: This breed is originated in Holland.

Distinguishing Characters

- Holsteins are large, stylish animals with color patterns of black and white or red and white.
- Holstein heifers can be bred at 15 months of age, when they weigh about 800 pounds. It is desirable to have Holstein females calve for the first time between 24 and 27 months of age.
- Milk yield - 7200-9000 kg.
- This is by far the best diary breed among exotic cattle regarding milk yield. On an average it gives 25 liter of milk per day, whereas a cross breed H.F. cow gives 10 - 15 liter per day.
- It can perform well in coastal and delta areas.

Holstein Friesian

Cross Breeding

It is mating of animals of different breeds. Cross breeding is followed for breeding animals for milk production and meat production. In India zebu breeds of cows and nondescript cows are crossed with exotic breeds like Holstein Friesian, Brown Swiss and Jersey bulls or their semen, to enhance the milk production potential of the progeny.

(*a*) As selection is a slow process of genetic improvement cross breeding has been taken up as the national breeding for improving milk production in India . Cross breeding word was initiated at NDRI Bangalore, Live Stock farm and Allahabad Agricultural Institute. At present cross breeding work is going on at Military dairy farms, NDRI Karnal, All India coordinated Research project son Cattle, Collaboration projects like Indo-Swiss, Indo Australian, Indo-Danish, projects and also in the field in farmer's he. The feeding and management of the crosses would be better, to enable them to express their production potential.

(*b*) In general the cross breeds were found to have higher birth weight, faster growth rate, earlier age at first calving, higher weight ; at first calving, higher lactation yield, longer lactation period) shorter service period, dry period and milk production and breeding efficiency.

(*c*) There are several exotic breeds being used in cross breeding programme, namely Holstein Friesian, Jersey, Brown Swiss and Reddane Holstein Friesian is found to be best suited for fluid milk supply in cities, and where higher feed inputs can be provided and where the temperature is temperate or sub-tropical. In contrast Jersey crosses are ideal when the milk is meant for product manufacture and where feed inputs are limited and the climate is trop.

Advantages

1. The desirable characters of the exotic parent are transmitted to the progeny which the indigenous parent does not have.

2. In India Cross-breeding and cows is done by using the exotic bulls and the progeny inherit the desirable characters of the parent like high milk yield early maturity, higher birth weight of calves, better growth rates, better reproductive efficiency and indigenous parents characters like, heat tolerance, disease resistance ability to thrive on scanty feeding and coarse fodder etc.

3. In pairs the way to evolve new breeds with desirable characters. Hybrid vigour is made use of in the progency.

4. Results are seen more quickly in characters like milk yield in the crossbred progeny.

Disadvantages

1. The breeding merit of cross breed animals may be slightly reduced.
2. Cross breeding requires maintenance of two or more pure breeds in order to product the cross breeds.

Cross-breed cattle

The crossbreeds are having exotic inheritance from Jersey, Brown Swiss or Holstein Friesian or a combination of these different breeds. Jersey breed is known for the milk fat percent and Holstein for the high quantity of milk.

Cross-breed cattle in India

S.No.	Name of the Breed	Native breed	Specific region	Assembling centre	Remarks
1	Brown Swiss	Switzerland	-	India, Pakistan & other Asiancountries	Dairy breed
2	Holstein Friesian	Holland	Province of North Holland and West Friesland	Throughout the country(crossbreds)	Dairy breed
3	Jersey	British Isles	Island of Jersey	Crossbreds available inall states.	Dairy breed

Indigenous cattle of India

S. No.	Breed	Habitat/ Main State	Breeding Tract District	Assembling Center	Areas of demand	Remarks
1.	Hallikar	Karnataka	Tumkur, Hassan & Mysore	Dodbalapur, Chickballapur, Harikar, Devargudda, Chikkuvalli, Karuvalli, Chittavadgi (T.N.) North Arcot (T.N.) Hindupur, Somaghatta, Anantpur (A.P.)	Dharwar, North Kanara, Bellary (KT) Anantur & Chittur (A.P.), Coimbatore North Arcot, Salem (T.N.)	Draught breed
2.	Kangayam	Tamil Nadu	Erode	Avanashi, Tirppur, Kannauram, Madurai Athicombu	Southern Districts of Tamil Nadu	Draught breed

Contd...

3.	Red Sindhi	Pakistan All parts of India	—	—	—	Dairy breed
4.	Tharparkar	Pakistan (sind)	Umarkot, Naukot, Dhoro Naro Chor	Balotra (Jodhpur), Puskar (Ajmer), Gujarat State	—	Dairy breed
5.	Vechur	Kerala	—	Vaikom, Mannuthy (Kerala State)	—	—

Different Economic Characters in Dairy Cattle?

The various economic characters in Dairy Cattle management are:

1. Lactation yield

2. Lactation period

3. Persistency of yield

4. Age at first calving

5. Service period

6. Dry period

7. Inter calving period

8. Reproductive efficiency

9. Efficiency of feed utilization

10. Disease resistance.

1. Lactation Yield

The lactation yield in a lactation period is known as lactation yield. 'The lactation yield in Indian breeds is very low compared to exotic breeds. This is dependent on of calving, frequency of milking, persistency of yield. Normally in dairy cattle 30 - 40% increase in milk production from first lactation to maturity is observed. After 3 or 4 lactation the production starts declining. For comparison of milk yield of different breeds and animals the milk yield should be converted into fat corrected milk (FCM). 4% FCM = 0.4 total milk + 15 total fat. After parturition the milk yield per day will be increased and reaches peak within 2-4 weeks after calving. This yield is known as peak yield. The maintenance of peak yield for more time is importance for better milk production. The lactation period in Indian breeds is low and so the production is also less and conversion.

2. Lactation Period

The length of milk producing period after calving is known as lactation period. The optimum lactation period is 305 days. The milk production will breed of dairy animals and farm records will be less, if this period is shortened. Indian breeds will have less lactation period, but in some breeds this period is more with very little milk production.

3. Persistency of Milk Yield

During lactation period the animal reaches maximum milk yield per day within 2-4 weeks which is called peak yield. For high level of lactation yield, this peak yield should be maintained for longer period as far as possible, The maintenance of peak yield for long period is known as persistency, slow decrease in dairy milk yield after reaching peak yield in necessary. High persistency is necessary to maintain high level of milk production.

4. Age at First Calving

The age o the animal at first calving is very important for high life time production. The desirable age at first calving in Indian breeds is 3 years, 2 years in cross breed cattle and 3 1/2 years in Buffaloes. Prolonged age at first calving will have high production in the first lactation) but the life time production will be decreased due to less no of calving. If the age at first calving is below optimum, the calves born are weak, difficulty in calving and less milk production in first lactation.

5. Service Period

It is the period between -date of calving and date of successful conception. The optimum service period helps the animal to recover from the stress of calving and also to get back the reproductive organs back to normal For cattle the optimum service period is 60-90 days. If the service period is too prolonged the calving interval prolonged, less no. of calving will be obtained in her life time and ultimately less life time production.' If the service period is too short, the animal will become weak and persistency of milk production is poor due to immediate pregnancy.

6. Dry Period

It is the period from the date of drying (stop of milk production) to next calving. When the animal in pregnancy, before next calving. The animal should be given rest period to compensate for growth of fetus. A minimum of 2 – 2 ½ months dry period should be allowed) If the dry period is not given or too low dry period, the animals suffer from stress and in next lactation, the milk production drops substantially and also it gives weak calves. On the other hand if the dry period given is too high, it may not have that much effect on increasing milk yield in the next lactation, but it decrease the production in the present lactation.

7. Inter Calving Period

This is the -period between two successive calving. It is more, profitable to have one calf yearly in cattle and at least one calf for every 15 months in buffaloes. If the calving interval is more, the total no. of carvings in her life time will be decreased and also total life production of milk decrease.

8. Reproductive Efficiency

The reproductive efficiency means the more number of calves during life time, so that total life time production is increased, The reproduction or breeding efficiency is determined by the combined effect of hereditary and environment. Several measures of breeding efficiency like number of services per conception, calving interval, and days from first breeding to conception are useful. Reproductive efficiency has generally a low heritability value indicating that most of the variations in this trait is due to non genetic factors. In adverse environmental conditions, the poor milk producing animals may not be much affected compared to high effect in high milk yield.

9. Efficiency of Feed Utilization and Conversion into Milk

The animal should take the feed more and utilize efficiently to convert into the milk.

10. Disease Resistance

Indian breeds are more resistant to majority of disease compared to exotic cattle. Cross breeding helps to get this character.

Different Breeds of Buffaloes in India

Sr.No.	Breed	Habitat/Main State	Breeding Tract Districts	Assembling Centers	Remarks
1.	Jafarabadi	Gujarat	Kathiawar and Honreli	Breeding areas of Saurashtra	Dairy breed
2.	Murrah	Haryana, Uttar Pradesh, Punjab	Rohtak, Hissar, Karnal, Jind, Gurgaon, Western parts of Uttar Pradesh Nabha and Patiala	Rohtak, Bahadurgarh, Delhi, Jahanzgarh, Mahim, Hissar, Bhiwani, Hansi, Rewari, Ferozpur, Jirka, Nangloi	Dairy Breed
3.	Surti	Gujarat	Kheda, Vadodara	Throughout Gujarat	Dairy breed

Brief Information about buffalo

The buffalo is a multipurpose animal. Not only is it a better source of milk than the cow, it also provides meat and works as a draught animal. Of all the domestic animals, the Asian buffalo holds the greatest promise and potential for production. It is well known that the buffalo is remarkable for its feed conversion ability. The production of buffalo milk in the Asian-Pacific region exceeds 45 million tonnes annually, of which over 30 million tonnes are produced in India alone. Buffaloes are labour intensive and cost-effective. They are the most versatile of all work animals in the variety of tasks, which they can be taught to undertake. All buffalo breeds have a strong milk/meat entity. When a buffalo is fed well and managed for early slaughter (at a live weight of 350 to 450 kg), a yield of palatable, high-grade meat can be obtained at a competitive cost.

Buffalo Breeds

1. Murrah

Origin: The home tract of this breed is mainly in Punjab and Delhi.

Distinguishing Characters

- The breed tract is Rohtak, Hisar and Jind of Haryana.
- The breed characteristics are massive body, neck and head comparatively long, horns short and tightly curled, Udder well developed, hip broad and fore – and hind quarters drooping.
- The average milk production per lactation is 1,500 to 2,500 kg.
- On an average the daily milk yield is found to be 8-10 liter, whereas a cross breed murrah buffalo gives 6-8 liter per day.
- The age at first calving is 45 – 50 months in villages but in good herds it is 36 – 40 months.
- The inter-calving period is 450 – 500 days
- It performs well in coastal and slightly cold climatic areas.

2. Surti

Origin: Gujarat.

Distinguishing Characters

- The Native tract of this breed is Kaira and Baroda districts of Gujarat.

- The body is well shaped and medium sized.
- The barrel is wedge shaped.
- The head is long with prominent eyes.
- The horns are sickle shaped, moderately long and flat.
- The colour is black or brown the peculiarity of breed is two white collars one round the jaw and the other at the brisket.
- The average milk yield is around 1700kgs.
- The age at first calving is 40 to 50 months with an intercalving period of 400 – 500 days.
- The bullocks are good for light work.

3. Jaffarabadi

- Kathiawar district of Gujarat
- They yield appreciable quantity of milk, with an exceptionally high butter fat content.
- The average milk yield is around 1800-2700 kg.

Indian Dairy Industry

Today, India is 'The Oyster' of the global dairy industry. It offers opportunities galore to entrepreneurs worldwide, who wish to capitalize on one of the world's largest and fastest growing markets for milk and milk products A bagful of 'pearls' awaits the international dairy processor in India. The Indian dairy industry is rapidly growing, trying to keep pace with the galloping progress around the world. As he expands his overseas operations to India many profitable options await him. He may transfer technology, sign joint ventures or use India as a sourcing center for regional exports. The liberalization of the Indian economy beckons to MNC's and foreign investors alike.

India's dairy sector is expected to triple its production in the next 10 years in view of expanding potential for export to Europe and the West. Moreover with WTO regulations expected to come into force in coming years all the developed countries which are among big exporters today would have to withdraw the support and subsidy to their domestic milk products sector. Also India today is the lowest cost producer of per liter of milk in the world, at 27 cents, compared with the U.S. 63 cents, and Japan's $2.8 dollars. Also to take advantage of this lowest cost of milk production and increasing production in the country multinational companies are planning to expand their activities here. Some of these milk producers have already obtained quality standard certificates from the authorities. This will help them in marketing their products in foreign countries in processed form. The urban market for milk products is expected to grow at an accelerated pace of around 33% per annum to around Rs. 43, 500 crores by year 2005. This growth is going to come from the greater emphasis on the processed foods sector and also by increase in the conversion of milk into milk products. By 2005, the value of Indian dairy produce is expected to be Rs 10, 00,000 million. Presently the market is valued at around Rs7, 00,000mn.

Background

India with 134mn cows and 125mn buffaloes has the largest population of cattle in the world. Total cattle population in the country as on October'00 stood at 313mn. More than fifty percent of the buffaloes and twenty percent of the cattle in the world are found in India and most of these are milch cows and milch buffaloes.

Indian dairy sector contributes the large share in agricultural gross domestic products. Presently there are around 70,000 village dairy cooperatives across the country. The co-operative societies are federated into 170 district milk producers unions, which is turn has 22-state cooperative dairy federation. Milk production gives employment to more than 72mn dairy farmers. In terms of total production, India is the leading producer of milk in the world followed by USA. The milk production in 1999-00 is estimated at 78mn MT as compared to 74.5mn MT in the previous year. This production is expected to increase to 81mn MT by 2000-01. Of this total produce of 78mn cows' milk constitute 36mn MT while rest is from other cattle. While world milk production declined by 2 per cent in the last three years, according to FAO estimates, Indian production has increased by 4 per cent. The milk production in India accounts for more than 13% of the total world output and 57% of total Asia's production. The top five milk producing nations in the world are India, USA, Russia, Germany and France.

Production of Milk in India

Year	Production in million MT
1988-89	48.4
1989-90	51.4
1990-91	53.7
1991-92	56.3
1992-93	58.6
1993-94	61.2
1994-95	63.5
1995-96	65
1996-97	68.5
1997-98	70.8
1998-99	74.7
1999-00 (E)	78.1
2000-01 (T)	81.0

E= estimated; T= target / expected

World's Major Milk Producers

Country	1997-98	1998-99 (Million MTs, Approx.)
India	71	74.5
USA	71	71
Russia	34	33
Germany	27	27
France	24	24
Pakistan	21	22
Brazil	21	27
UK	14	14
Ukraine	15	14
Poland	12	12
New Zealand	11	12
Netherlands	11	11
Italy	10	10
Australia	9	10

Operation Flood

The transition of the Indian milk industry from a situation of net import to that of surplus has been led by the efforts of National Dairy Development Board's Operation Flood. programme under the aegis of the former Chairman of the board Dr. Kurien. Launched in 1970, Operation Flood has led to the modernization of India's dairy sector and created a strong network for procurement processing and distribution of milk by the co-operative sector. Per capita availability of milk has increased from 132 gm per day in 1950 to over 220 gm per day in 1998. The main thrust of Operation Flood was to organize dairy cooperatives in the milkshed areas of the village, and to link them to the four Metro cities, which are the main markets for milk. The efforts undertaken by NDDB have not only led to enhanced production, improvement in methods of processing and development of a strong marketing network, but have also led to the emergence of dairying as an important source of employment and income generation in the rural areas. It has also led to an improvement in yields, longer lactation periods, shorter calving intervals, etc through the use of modern breeding techniques. Establishment of milk collection centers and chilling centers has enhanced life of raw milk and enabled minimization of wastage due to spoilage of milk. Operation Flood has been one of the world's largest dairy development programme and looking at the success achieved in India by adopting the co-operative route, a few other countries have also replicated the model of India's White Revolution.

Per Capita Availability of Milk

Year	gm/day
1950	132
1960	127
1968	113
1973	111
1980*	128
1990	178
1992	192
1996	198
1997	200
1998	202
1999	203
2000	212
2001E	225
2002P	250

E=Estimated; P= Provisional

What does the Indian Dairy Industry has to Offer to Foreign Investors?

India is a land of opportunity for investors looking for new and expanding markets. Dairy food processing holds immense potential for high returns. Growth prospects in the dairy food sector are termed healthy, according to various studies on the subject.

The basic infrastructural elements for a successful enterprise are in place.

- Key elements of free market system
- raw material (milk) availability
- an established infrastructure of technology
- supporting manpower

An entrepreneur's participation is likely to provide attractive returns on the investment in a fast growing market such as India, along with an export potential in the Middle East, Singapore, Malaysia, Indonesia, Korea, Thailand, Hong Kong and other countries in the region. Among several areas of potential participation by NRIs and foreign investors, the following list outlines a few promising opportunities:

Where biotechnology can use in dairy industry?

- Dairy cattle breeding of the finest buffaloes and hybrid cows

- Milk yield increase with recombinant somatotropin

- Recombinant chymosin, acceptable to vegetarian consumers

- Dairy cultures, probiotics, dairy biologics, enzymes and coloring materials for food processing

- Fermentation derived foods and industrial products alcohol, citric acid, lysine, flavor preparations, etc.

- Biopreservative ingredients based on dairy fermentation, viz., Nisin, pediococcin, acidophilin, and bulgarican contained in dairy powders.

FUTURE PROSPECTS FOR INDIAN DAIRY

India is the world's highest milk producer and all set to become the world's largest food factory. In celebration, Indian Dairy sector is now ready to invite NRIs and Foreign investors to find this country a place for the mammoth investment projects. Be it investors, researchers, entrepreneurs, or the merely curious – Indian Dairy sector has something for everyone. Milk production is relatively efficient way of converting vegetable material into animal food. Dairy cows' buffalo's goats and sheep can eat fodder and crop by products which are not eaten by humans. Yet the loss of nutrients energy and equipment required in milk handling inevitably make milk comparatively expensive food. Also if dairying is to play its part in rural development policies, the price to milk producers has to be remunerative. In a situation of increased international prices, low availabilities of food aid and foreign exchange constraints, large scale subsidization of milk conception will be difficult in the majority of developing countries.

Hence in the foreseeable future, in most of developing countries milk and milk products will not play the same roll in nutrition as in the affluent societies of developed countries. Effective demand will come mainly from middle and high income consumers in urban areas. There are ways to mitigate the effects of unequal distribution of incomes. In Cuba where the Government attaches high priority to milk in its food and nutrition policy, all pre-school children receive a daily ration of almost a liter of milk fat the reduced price. Cheap milk and milk products are made available to certain other vulnerable groups, by milk products outside the rationing system are sold price which is well above the cost level. Until recently, most fresh milk in the big cities of China was a reserved for infants and hospitals, but with the increase in supply, rationing has been relaxed. In other countries dairy industries have attempted to reach lower income consumers by variation of compositional quality or packaging and distribution methods or blending milk in vegetable ingredients in formula foods for vulnerable groups. For instance, pricing of products rich in butter fat or in more luxury packaging above cost level so as to enable sales of high protein milk products at a somewhat a reduced price has been widely practiced in developing countries. This policies need to be brought in Indian Dairy scenario.

What is mean by Cooperative Dairy Farming?

District Cooperative Milk Producer's Union (Dugdh Sangh)

Main functions of the union are:

- Procurement of milk from the village milking societies of the district,

- Arranging transportation of raw milk from the VDCS to the Milk Union,

- Providing input services to the producers like veterinary care, artificial insemination services, cattle-feed sales, mineral mixture sales, fodder and fodder seed sales,

- Conducting training on cooperative development, animal husbandry and dairying for milk producers and conducting skill development and leadership development training for VDCS staff and Management Committee members,

- Providing management support to the VDCS along with supervision of its activities.

- Establish chilling centres and dairy plants for processing the milk received from the villages.

- Selling liquid milk and milk products within the district

- Process milk into milk products as per the requirement of State Marketing Federation.

- Decide on the prices of milk to be paid to milk producers as well on the prices of support services provided to members.

State Cooperative Milk Federation (Federation)

The main functions of the federation are as follows:

- Marketing of milk and milk products processed/manufactured by Milk Unions,

- Establish a distribution network for marketing of milk and milk products,

- Arranging transportation of milk and milk products from the Milk Unions to the market,

- Creating and maintaining a brand for marketing of milk & milk products,

- Providing support services to the Milk Unions and members like technical inputs, management support and advisory services,

- Pooling surplus milk from the Milk Unions and supplying it to deficit Milk Unions,

- Establish feeder-balancing dairy plants for processing the surplus milk of the Milk Unions,

- Arranging for common purchase of raw materials used in manufacture/ packaging of milk products,

- Decide on the prices of milk and milk products to be paid to Milk Unions,

- Decide on the products to be manufactured at Milk Unions and capacity required for the same.

- Conduct long-term milk production, procurement and processing as well as marketing planning.

- Arranging finance for the Milk Unions and providing them technical know-how.

- Designing and providing training in cooperative development and technical and marketing functions.

- Conflict resolution and keeping the entire structure intact.

Today, there are around 176 cooperative dairy unions formed by 125,000 dairy cooperative societies, having a total membership of around 13 million farmers on the same pattern, who are processing and marketing milk and milk products profitably, be it Amul in Gujarat or Verka in Punjab, Vijaya in Andhra Pradesh, Milma in Kerala, Gokul in Maharashtra, Saras in Rajasthan or a Nandini in Karnataka. This process has created more than 190 dairy processing plants spread all over India with large investments by these farmers' institutions. These cooperatives today collect approximately 23 million kg of milk per day and pay an aggregate amount of more than Rs. 125 billion to the milk producers in a year.

Cattle Population in World

Country	Population
India	281,700,000
Brazil	187,087,000
China	139,721,000
USA	96,669,000
EU	87,650,000
Argentina	51,062,000
Pakistan	38,300,000
Australia	29,202,000
Mexico	26,489,000
Russia Federation	18,370,000
South Africa	14,187,000
Canada	13,945,000
Other	49,756,000

Livestock Farming

Livestock are *domesticated animals* raised in an agricultural setting to produce commodities such as food, *fiber* and *labor*. This article does not discuss *poultry* or farmed *fish*, although these, especially poultry, are commonly included within the meaning of "livestock". Livestock are generally raised for profit. Raising animals (*animal husbandry*) is a component of modern *agriculture*. It has been practiced in many cultures since the transition to *farming* from *hunter-gather* lifestyles.

The term "livestock" is nebulous and may be defined narrowly or broadly. On a broader view, livestock refers to any breed or population of animal kept by humans for a useful, commercial purpose. This can mean *domestic animals, semi-*domestic animals, or *captive wild animals*. Semi-domesticated refers to animals which are only lightly domesticated or of disputed status. These populations may also be in the process of *domestication*. Some people may use the term livestock to refer to only domestic animals or even to only *red meat* animals.

Animal / Type	Domestication status	Time of first captivity, domestication	Area of first captivity, domestication	Current commercial uses
Alpaca Mammal, herbivore	domestic	Between 5000 BC and 4000 BC	Andes	wool, meat
Banteng Mammal, herbivore	domestic	Unknown Java	Southeast Asia,	meat, milk, draught
Bison Mammal, herbivore	captive (see also Beefalo)	Late 19th Century	North America	meat, leather
Camel Mammal, herbivore	domestic	Between 4000 BC and 1400 BC	Asia	mount, pack animal, meat, dairy, camel hair
Cat Mammal, carnivore	domestic	7500 BC	Near East	pest control, companionship, meat
Cattle Mammal, herbivore	domestic	6000 BC	Southwest Asia, India, North Africa	Meat (beef, veal, blood), dairy, leather, draught
Deer Mammal, herbivore	captive	1st century AD	UK	Meat (venison), leather, antlers, antler velvet
Dog Mammal, omnivore	domestic	12000 BC		pack animal, draught, hunting, herding, searching/ gathering, watching/ guarding, meat
Donkey Mammal, herbivore	domestic	4000 BC	Egypt	mount, pack animal, draught, meat, dairy

Contd...

Animal / Type	Domestication status	Time of first captivity, domestication	Area of first captivity, domestication	Current commercial uses
Gayal Mammal, herbivore	domestic	Unknown	Southeast Asia	meat, draught
Goat Mammal, herbivore	domestic	8000 BC	Southwest Asia	Dairy, meat, wool, leather, light draught
Guinea pig Mammal, herbivore	domestic	5000 BC	South America	Meat
Horse Mammal, herbivore	domestic	4000 BC	Eurasian Steppes	Mount, draught, dairy, meat, pack animal
Llama Mammal, herbivore	domestic	3500 BC	Andes	light mount, pack animal, draught, meat, wool
Mule Mammal, herbivore	domestic			mount, pack animal, draught
Pig Mammal, omnivore	domestic	7000 BC	Eastern Anatolia	Meat (pork, bacon, etc.), leather, pet, research
Rabbit Mammal, herbivore	domestic	between AD 400-900	France	Meat, fur, leather pet, research
Reindeer Mammal, herbivore	semi-domestic	3000 BC	Northern Russia	Meat, leather, antlers, dairy, draught
Sheep Mammal, herbivore	domestic	Between 11000 BC-9000 BC	Southwest Asia	Wool, dairy, leather, meat (mutton, lamb)
Water buffalo Mammal, herbivore	domestic	4000 BC	South Asia	mount, draught, meat, dairy
Yak Mammal, herbivore	domestic	2500 BC	Tibet, Nepal	Meat, dairy, wool, mount, pack animal, draught

PROCEDURE FOR SILAGE PREPARATION

Silage is *fermented*, high-moisture stored *fodder* which can be fed to *ruminants* (*cud*-chewing animals such as *cattle* and *sheep*) or used as a *biofuelfeedstock* for *anaerobic digesters*. It is fermented and stored in a process called *ensilage, ensiling* or *silaging*, and is usually made from *grass* crops including *maize, sorghum* or other *cereals*, using the entire green plant (not just the grain). Silage can be made from many field crops, and special terms may be used depending on type (*oatlage* for oats, *haylage* for *alfalfa*, but see *below* for the different British use of the term *haylage*). Silage is made either by placing cut green vegetation in a *silo*, by piling it in a large heap covered with plastic sheet, or by wrapping large bales in plastic film.

The crops suitable for ensilage are the ordinary grasses, *clovers, alfalfa, vetches, oats, rye* and *maize*; various *weeds* may also be stored in silos, notably *spurrey* such as *Spergula arvensis*. Silage must be made from plant material with suitable moisture content, about 50% to 60% depending on the means of storage, the degree of compression, and the amount of water that will be lost in storage, but not exceeding 75%. Weather during harvest need not be as fair and dry as when harvesting for drying. For corn, harvest begins when the whole-plant moisture is at a suitable level, ideally a few days before it is ripe. For pasture-type crops, the grass is mowed and allowed to wilt for a day or so until the moisture content drops to a suitable level. Ideally the crop is mowed when in full *flower,* and deposited in the silo on the day of its cutting. After harvesting, crops are shredded to pieces about 0.5 in (1.3 cm) long. The material is spread in uniform layers over the floor of the silo, and closely packed. When the silo is filled or the stack built, a layer of *straw* or some other dry porous substance may be spread over the surface. In the silo the pressure of the material, when *chaffed,* excludes air from all but the top layer; in the case of the stack extra *pressure* is applied by weights in order to prevent excessive heating.

Forage harvesters collect and chop the plant material, and deposit it in trucks or wagons. These forage harvesters can be either tractor-drawn or self-propelled. Harvesters blow the silage into the wagon via a chute at the rear or side of the machine. Silage may also be emptied into a bagger, which puts the silage into a large plastic bag that is laid out on the ground. In North America, Australia, northwestern Europe, and frequently in New Zealand, silage is placed in large heaps on the ground and rolled by tractor to push out the air, then wrapped in plastic covers held down by reused tires or tire ring walls. In New Zealand and Northern Europe, the silo or "pit" is often a bunker built into the side of a bank, usually made out of concrete or old wooden railroad ties (railway sleepers). The chopped grass can then be dumped in at the top, to be drawn from the bottom in winter. This requires considerable effort to compress the stack in the silo to cure it properly. Again, the pit is covered with plastic sheet and weighed down with tire weights.

In an alternative method, the cut vegetation is baled, making *balage* (North America) or *silage bales.* The grass or other forage is cut and partly dried until it contains 30–40% moisture (much drier than bulk silage, but too damp to be stored as dry hay). It is then made into large bales which are wrapped tightly in plastic to exclude air. The plastic may wrap the whole of each cylindrical or cuboid bale, or be wrapped around only the curved sides of a cylindrical bale, leaving the ends uncovered. In this case, the bales are placed tightly end to end on the ground, making a long continuous "sausage" of silage, often at the side of a field. The wrapping may be performed by a bale wrapper, while the baled silage is handled using a bale handler or a front-loader, either impaling the bale on a flap, or by using a special grab. The flaps do not hole the bales.

In the UK, baled silage is most often made in round bales about 4 feet by 4 feet, individually wrapped with four to six layers of "bale wrap plastic" (black, white or green 25 micro meter stretch film). The dry matter can vary a lot but can be from about 20% dry matter upwards. The continuous "sausage" referred to above is made with a special machine which wraps the bales as they are pushed through a rotating hoop which applies the bale wrap to the outside of the bales (round or square) in a continuous wrap. The machine places the bales on the ground after wrapping by moving forward slowly during the wrapping process (search for "tube liner" various makes). Haylage is a name for high dry matter silage of around 45% to 75%. Horse haylage is usually 55% to 75% dry matter, made in small bales or larger bales. Handling of wrapped bales is most often with some type of gripper that squeezes the plastic-covered bale between two metal parts to avoid puncturing the plastic. Simple fixed versions are available for round bales which are made of two shaped pipes or tubes spaced apart to slide under the sides of the bale, but when lifted will not let it slip through. Often used on the tractor rear three-point linkage, they incorporate a trip tipping mechanism which can flip the bales over on to the flat side/end for storage on the thickest plastic layers.

Silage undergoes *anaerobic* fermentation, which starts about 48 hours after the silo is filled, and converts sugars to acids. Fermentation is essentially complete after about two weeks. Before anaerobic fermentation starts, there is an aerobic phase in which the trapped oxygen is consumed. The closeness with which the fodder is packed, determines the nature of the resulting silage by regulating the *chemical reactions* that occur in the stack. When closely packed, the supply of *oxygen* is limited; and the attendant *acid fermentation* brings about the decomposition of the *carbohydrates* present into *acetic, butyric* and *lactic acids*. This product is named sour silage. If, on the other hand, the fodder is unchaffed and loosely packed, or the silo is built gradually, *oxidation* proceeds more rapidly and the *temperature* rises; if the mass is compressed when the temperature is 140 to 160 Fahrenheit, the action ceases and sweet silage results. The *nitrogenous* ingredients of the fodder also suffer change: in making sour silage as much as one-third of the *albuminoids* may be converted into *amino* and *ammonium* compounds; while in making sweet silage a smaller proportion is changed, but they become less *digestible*. If the fermentation process is poorly managed, sour silage acquires an unpleasant *odour* due to excess production of ammonia or butyric acid (the latter is responsible for the smell of rancid butter). In the past, the fermentation was conducted by indigenous microorganisms, but, today, some bulk silage is inoculated with specific microorganisms to speed fermentation or improve the resulting silage. Silage inoculants contain one or more strains of *lactic acid bacteria*, and the most common is *Lactobacillus plantarum*. Other bacteria used in inoculants include Lactobacillus buchneri, Enterococcus faecium and Pediococcus species.

Silage must be firmly packed to minimize the oxygen content, or it will spoil. Silage goes through four major stages in a silo:

- Presealing, which, after the first few days after filling a silo, enables some respiration and some dry matter (DM) loss, but stops

- Fermentation, which occurs over a few weeks; pH drops; there is more DM loss, but hemicellulose is broken down; aerobic respiration stops

- Infiltration, which enables some oxygen infiltration, allowing for limited microbial respiration; available carbohydrates (CHOs) are lost as heat and gas

- Emptying, which exposes surface, causing additional loss; rate of loss increases.

Ensilage can be substituted for root crops. Bulk silage is commonly fed to dairy cattle, while baled silage tends to be used for beef cattle, sheep and horses. The advantages of silage as animal feed are several:

- During fermentation, the silage bacteria act on the cellulose and carbohydrates in the forage to produce volatile fatty acids (VFAs), such as acetic, propionic, lactic, and butyric acids. By lowering pH, these create a hostile environment for competing bacteria that might cause spoilage. The VFAs thus act as natural preservatives, in the same way that the lactic acid in yogurt and cheese increases the preserve ability of what began as milk, or vinegar (dilute acetic acid) preserves pickled vegetables. This preservative action is particularly important during winter in temperate regions, when green forage is unavailable.

- When silage is prepared under optimal conditions, the modest acidity also has the effect of improving palatability and provides a dietary contrast for the animal. (However, excessive production of acetic and butyric acids can reduce palatability: the mix of bacteria is ideally chosen so as to maximize lactic acid production.

- Several of the fermenting organisms produce vitamins: for example, lactobacillus species produce folic acid and vitamin B12

- The fermentation process that produces VFA also yields energy that the bacteria use: some of the energy is released as heat. Silage is thus modestly lower in caloric content than the original forage, in the same way that yoghurt has modestly fewer calories than milk. However, this loss of energy is offset by the preservation characteristics and improved digestibility of silage.

SHEEP AND GOATS

Different Indian sheep breeds?

1. Mecheri

- It is distributed in Salem, Erode, Karur, Namakkal, and fewer parts of Dharmapuri districts of Tamilnadu.

- It is a meat purpose breed.
- It has medium sized body with pale purplish skin color.
- There are no horns for both the sexes.
- Tail is smaller and slender.
- Adult male average body weight 36kg.
- Adult female average body weight 22kg.

2. Chennai red

- This is distributed in Chennai, Kancheepuram, Villupuram, Thiruvannamalai districts of Tamilnadu.
- It is meat purpose breed.
- Majority are purple in color.
- Certain animals have colored stripes on their forehead.
- Adult male average body weight 36kg.
- Adult females average body weight 24 kg.

3. Ramanadhapuram white

- This is distributed in Ramanadhapuram, Sivagangai, and Virudhunagar districts of Tamilnadu.
- It is meat purpose breed.
- It has medium sized body.
- Majority of them are white in color.
- Certain goats hold black colored stripes all over their body.
- Adult males have their bent horns, whereas females with absence of horns.
- Legs are smaller and slender.
- Adult male average body weight 31kg.
- Adult female average body weight 23kg.

4. Keezhakaraisal

- This is distributed in Ramanadhapuram, Sivagangai and Thirunelveli districts of Tamilnadu.
- It is meat purpose breed.

- It has medium sized body.
- It is found in black red skin color.
- Black colored bands are found on the skin in the regions of head, stomach and legs.
- Tail is smaller and slender.
- Adult male goats are found with stronger coiled horns.
- Majority of the goats are found with wattle, under the jaw/throat.
- Adult male average body weight 29kg.
- Adult female average body weight 22kg.

5. Vembur

- It is distributed in Vembur, melakarandhai, keezha karandhai, nagalapuram regions, Tuticorin and Virudhunagar districts of Tamilnadu.
- It is meat purpose breed.
- These are taller breeds.
- They have white color skin with red color spots over their body.
- Ears are drooped out.
- Tail is smaller and slender.
- Adult males are found with horns and absence of horns in case of females.
- Adult males average body weight 35kgs.
- Adult females average body weight 28kgs.

6. Neelagiri

- These are distributed in Neelagiri district of Tamilnadu.
- It is wool purpose breed.
- They are medium weighed animal.
- Majority are found in white colors.
- Certain goats are found with purple spots on their body and face.
- Ears are broad and drooped out.
- Females are without horns.
- Adult male average body weight 31kg.
- Adult female average body weight 31kg.

7. Trichy black

- These are distributed in Trichy, Perambalur, Dharmapuri and Salem districts of Tamilnadu.
- It is wool purpose breed.
- These are smaller breeds.
- Black coloured all over the body.
- Adult males are found with horns and females without horns.
- Ears are smaller, facing forward and downwards.
- Adult male average body weight 26kg.
- Adult female average body weight 19kg.

8. Coimbatore

- It is distributed in Coimbatore district of Tamilnadu.
- It is wool purpose breed.
- Medium weighed animal.
- Found in white colors, with black or purple colored bands, seen over the regions of head and neck.
- 30% of adult females are free of horns.
- Adult male average body weight 25kg.
- Adult female average body weight 20kg.

9. Deccani

- Deccani breed is an admixture of the woolly types of Rajasthan and the hairy types of Andra Pradesh and Tamil Nadu.
- It is found in Bombay-Deccan region and parts of Karnataka and Andhra Pradesh States.
- The sheep is small and hardy, and well adapted to poor pastoral conditions.
- It possesses a coloured fleece, black and gray colours being more dominant.
- The average annual yield of wool being 4.54 kg per sheep.
- The wool is of a low grade and is a mixture of hair and fine fibres, mostly consumed for the manufacture of rough blankets (Kambals).
- The flocks are maintained chiefly for mutton.

10. Nellore

- It is distributed in Nellore, Prakasam and Ongole districts of Andhra Pradesh.
- They are tall animals with little hair except at brisket, withers and breech.
- Rams are homed ewes are polled.
- Long and drooping ears;
- Majority of animals carry wattles.
- Males have average body weight of 36 kg and female have 28 kg.
- Nellore is the tallest breed of sheep in India, resembling goats in appearance.
- It has a long face and long ears with the body densely covered with short hair.
- The majority of the flocks are of fawn or deep red fawn colour.

11. Mandya

- It is distributed in Mandya district of Karnataka.
- Relatively small animals colour white - sometimes face is light brown which may extend up to neck.
- Compact body with typical reversed "U" shaped conformation from the rear.
- Ears long, leafy and drooping.
- Both sexes polled.
- Coat extremely coarse and hairy adult male weighs 35 kg and female weighs 23 kg.
- Best mutton type conformation among the Indian breeds.

12. Marwari

- Sheep are hardy, yielding coarser carpet variety white wool of a mixed hairy composition.
- This sheep is characterized by long legs, black face and a prominent nose.
- Fleshy appendages under throat, known as wattles, are often present.
- Tail is short and pointed.
- The sheep are found all over Jodhpur and parts of Jaipur districts.

- Flocks are raised in Pali and Barmer districts.
- The animals migrate to distant places in Uttar Pradesh, remote districts of Madhya Pradesh and sometimes to the northern parts of Maharashtra.
- They possess high resistance to disease and worms.
- The yield of wool per year is 0.90-1.81 kg per animal.

13. Gaddi

- Sheep are small in size, and are found in Kishtwar and Bhadarwah tehsils of Jammu.
- A large number inhabit the Kulu valleys in HP winter, and in summer they graze the highest elevations of them Pir Panjal Mountains, mostly in the Paddar range.
- Rams are horned, ewes hornless; fleece is generally white with brown coloured hair on the face.
- Wool is fine and lustrous; average annual yield is 1.13 kg per sheep, clipped thrice a year.
- A part of this clip is sent to Dhariwal mills and Amritsar markets.
- Undercoat is used for the manufacture of high quality Kulu shawls and blankets.

EXOTIC SHEEP BREEDS

1. Dorset

- This is native of U.K and are polled and horned.
- Face, ears and legs white in colour and free from wool.
- Wool yields 2.75 to 3.25 kg produce mutton of superior quality.
- Rams weight 80-110 kg and ewes weigh - 50 to 80 kg.
- It is a hardy breed and capable of performing well under most conditions.

2. Suffolk

- It is native of U.K. and is large animals with black face, ears and legs.
- Head and ears entirely free from wool.
- Both rams and ewes are polled though rams sometimes have scurs.
- Its average wool yield 2-3 kg.
- Mature Rams weigh 100-135 kg and ewes from 70-100 kg.

- Ewes are very prolific and excellent milkers.
- Suffolks imported to India have performed poorly than Dorsets.

3. Merino

- The most popular fine wool breed of the world, originated in Spain.
- It is a white faced sheep with white feet.
- Rams have horns whereas the ewes are hornless.
- Most of the head and legs are covered by wool.
- The animal is extremely hardy being able to survive under adverse weather and poor grazing conditions.
- The ewes live and yield longer than any other breed.

4. Rambouillet

- It was developed in France.
- This breed has a large head with white hair around the nose and ears.
- Rams have horns and ewes are hornless.
- Rams weigh as much as 125 kg and ewes up to a maximum of 90 kg.
- It produces an excellent fine-wool fleece.
- The fleece is heavy, close, compact, covering most of the body including face and legs.

5. Cheviot

- Is a medium wool breed, primarily developed in Scotland?
- The breed is small with erect ears, a clean white face and white legs, covered with short white hair.
- The nose, lips and feet are black.
- Rams weigh on an average up to 80 kg and ewes up to 55kg.

6. Southdown

- This breed is one of the oldest English breeds and has greatly contributed to the development of many other breeds of sheep.
- It is a small sheep excellent for mutton production.
- Body is low-set, compact, wide and deep with legs wide apart. Head is broad; face is light brown coloured.
- Rams weigh about 80 kg and ewes 55kg at maturity.

DIFFERENT INDIAN GOAT BREEDS

1. Jamunapari

- Jamunapari breeds are found mainly in the state of Uttar Pradesh.
- Its coat colour is white with tan or black markings at neck and ears.
- They are beard in both sexes; have tuft of long hairs in the buttocks.
- It is largest and most elegant of the long-legged goats of India.
- It has pronounced Roman nose having a tuft of hair which results in parrot mouth appearance.
- Their horns are short and flat and horizontally twisting backward.
- An adult male ranges from 90 to 100 cm in height, whereas a female goat ranges from 70 to 80 cm in height.
- It is, tall and leggy with convex face line and large folded pendulous ears.
- Generally found in white colors.
- Their ears are large and drooped downwards.
- An adult female weighs between 45kgs to 60kgs, whereas an adult male ranges between 65kgs to 80kgs.
- Average birth weight is up to 4 kg.
- Average age at first *kidding* is 20-25 months.
- They have large udder and big teats and average yield is 280 kg/274 days.
- Have the ability to yield 2 to 2.5kgs of milk per day.
- The fat content of the milk ranges between 3 to 3.5%.
- They thrive best under range conditions with plenty of shrubs for browsing.

2. Beetal

- It is are found mainly in the state of Punjab
- These breeds are grown mainly for the purpose of milk and meat.
- Generally smaller than the breed of Jamunapari.
- Coat Colour is predominantly black or brown with white spots of distorting size
- Males usually possess beard.
- They are not so heavy in weight.

- Average birth weight - 3 kg.
- An adult female goat ranges between 40kgs to 50kgs, whereas an adult male ranges between 50kgs to 70kgs.
- Age at first kidding - 20-22 months.
- Average lactation yield - 150 kg.
- They are having the ability to give, one kg to two kg of milk per day.
- Maximum yield being 591.5 kg in a lactation period of 177 days.

3. Barbari

- This is short haired and erect-horned goat popular in urban areas of Delhi, Uttar Pradesh, Gurgaon, Karnal, Panipat and Rohtak in Haryana state.
- Barbari breeds are grown mainly for milk and meat purpose.
- The color of this breed is white with light brown patches.
- An adult female goat weighs between 25kgs to 35kgs, whereas an adult male goat ranges between 35kgs to 45kgs.
- They are having the ability to give one kg to 1.5kgs of milk per day.
- These breeds have better reproductive capabilities.
- They will give, 2 to 3 kids in parturition.
- They are usually stall-fed and are reported to yield 0.90-1.25 kg of milk (fat content 5%) a day in a lactation period of 108 days
- They are prolific breeder and kid twice in 12-15 months.

4. Tellicherry

- Tellicherry breed is also called as malabari breed.
- It is found mostly in the state of Kerala.
- It is grown mostly for the purpose of meat.
- Generally seen in white, purple and black colors.
- An adult female ranges in weight from 30 to 40kgs, whereas an adult male ranges between 40 to 50kgs.
- They can yield one kg to two kg of milk per day.
- These types of breeds have better reproductive capabilities.
- They can give two to three kids in parturition.

5. Sirohi

- Coat colour is brown, white, and admixture of colours in typical patches; hair coarse and short.
- Compact and medium sized body.
- Tail twisted and carries coarse pointed hair.
- Horns are small and pointed, curved upward and backward.
- Average body weight of buck is 50 and doe is 23 kg.
- Average birth weight is 2.0 kg.
- Kidding is once a year, twins are common.
- Average age at first kidding is 19 months.
- Average lactation yield - 71 kg.
- Average lactation length - 175 days.

6. Osmanabadi

- Coat colour is predominantly black; white, brown and spotted occur.
- Long and short-haired type, based on presence or absence of long hair on the thighs and hind quarters.
- Tall and large size body and legs.
- Average birth weight 2.4 kg.
- Kidding is once a year.
- Average age at first kidding 19-20 months.
- It has good quality meat.
- Good yielders produce up to 3.5 kg a day.
- Average milk yield 170-180 kg per lactation.

7. Kanni aadu

- These are the tallest goat breeds found in Thirunelveli and Ramanadhapuram districts of Tamilnadu.
- Black or white spots in the black background are the characteristics colors of this breed.
- They are usually grown for meat purpose.
- The adult females of this breed ranges from 25kgs to 30kgs and the adult males' ranges from 35kgs to 40kgs in body weight.

- They are having ability to give birth to 2 to 3 kids.
- They grow well in the draught regions.

8. Kodi aadu

- These breeds are taller and found with different colors, but predominantly black.
- They usually give birth to one or two kids.
- They are usually grown for the purpose of guiding the goat flocks, which goes for grazing.
- These types of breeds are mostly found in the districts of Sivagangai, Ramanadhapuram, and Tuticorin districts of Tamilnadu.

9. Black Bengal

- Coat colour is predominantly black, brown/grey and white with soft, glossy and short hairs.
- Dwarf in body size, legs short, straight back; both sexes are bearded.
- Average live weight of buck is 15 kg and doe is 12 kg.
- Most prolific among the Indian breeds.
- Multiple births are common - two, three or four kids are born at a time.
- Kidding is twice a year. Average litter size is 2.1.
- Average age at first kidding is 9-10 months.
- Average lactation yield is 53 kg. Lactation length is 90 to 120 days.
- Its skin is in great demand for high quality shoe-making.

10. Chegu

- Coat colour is predominantly white but grayish red and mixed colours are also seen.
- Average buck live weight of buck is 39 kg and doe is 26 kg.
- Average birth weight is 2.0 kg.
- Kidding is once a year and mostly single.
- Average lactation yield is 69 kg and lactation length is 187 days.
- Used for draught to carry salt and small loads.
- Have long hair with under coat of delicate fibre below (cashmere or pashm).

- Legs are medium sized. Face and muzzle is tapering. Ears are Small.
- Horns are bent upward, backward and outward with one or more twists.
- Used for draught (pack) to carry salt and small loads.

11. *Changthangi*

- Predominantly white and the rest are brown, grey and black. Undercoat white/grey; yields warm delicate fibre - pashmina (cashmere, pashm.
- Body and legs are small, have strong body and powerful legs.
- Ears are small, pricked and pointed outwards.
- Horns are large turning outward, upward and inward forming a semicircular ring.
- Average live weight of buck is 20 and doe is 20 kg; average birth weight is 2.1 kg.
- Kidding is once a year, normally single;
- Average age at first kidding is 20 months.

WORLD GOAT POPULATION :
SELECTED REGIONS AND COUNTRIES, 2008

World sheep Population : Selected Regions and Countries, 2008

Country	Global sheep stocks in 2008, (Million)
China	136.4
Australia	79.0
India	65.0
Iran	53.8
Sudan	51.1
New Zealand	34.1
Nigeria	33.9
United Kingdom	33.1
World Total	1,078.2

Country/Region	Total Animals (millions)	Goat Milk (MT)	Goat Meat (million MT)
World	–	15.2	4.8
Africa	294.5	3.2	1.1
Nigeria	53.8	N/A	0.26
Sudan	43.1	1.47	0.19
Asia	511.3	8.89	3.4

Contd.

Country/Region	Total Animals (millions)	Goat Milk (MT)	Goat Meat (million MT)
Afghanistan	6.38	0.11	0.04
Pakistan	60.00	N/A	N/A
India	125.7	4.0	0.48
Bangladesh	56.4	2.16	0.21
China	149.37	0.26	1.83
Saudi Arabia	2.2	0.076	0.024
Americas	37.3	0.54	0.15
Mexico	8.8	0.16	0.04
USA	3.1	N/A	0.022
Europe	17.86	2.59	0.012
UK	0.09	N/A	N/A
France	1.2	0.58	0.007
Oceania	3.42	0.0004	0.018

Milk and Milk Products

A dairy product is food produced from the milk of mammals. Dairy products are usually high energy-yielding food products. A production plant for the processing of milk is called a dairy or a dairy factory. Apart from breastfed infants, the human consumption of dairy products is sourced primarily from the milk of cows, yet goats, sheep, yaks, horses, camels, and other mammals are other sources of dairy products consumed by humans. Dairy products are commonly found in European, Middle Eastern, and Indian cuisine, whereas aside from Mongolian cuisine they are little-known in traditional Asian cuisine.

♦ *Milk* after optional *homogenization, pasteurization,* in several grades after standardization of the fat level, and possible addition of bacteria *Streptococcus lactis* and *Leuconostoc citrovorum*

- *Crème fraîche,* slightly fermented cream

- *Clotted cream,* thick, spoonable cream made by heating

- *Smetana,* Central and Eastern European variety of sour cream

- *Cultured milk* resembling buttermilk, but uses different yeast and bacterial cultures

- *Kefir,* fermented milk drink from the *Northern Caucasus*

- *Kumis/Airag,* slightly fermented mares' milk popular in *Central Asia*

- Powdered milk (or milk powder), produced by removing the water from (usually skim) milk

- Whole milk products
- Buttermilk products
- Skim milk
- Whey products
- Ice cream
- High milk-fat and nutritional products (for infant formulas)
- Cultured and confectionery products
- Condensed milk, milk which has been concentrated by evaporation, with sugar added for reduced process time and longer life in an opened can
- Khoa, milk which has been completely concentrated by evaporation, used in Indian cuisine including gulab jamun, peda, etc.)
- Evaporated milk, (less concentrated than condensed) milk without added sugar
- Ricotta, acidified whey, reduced in volume
- Infant formula, dried milk powder with specific additives for feeding human infants
- Baked milk, a variety of boiled milk that has been particularly popular in Russia

◆ Butter, mostly milk fat, produced by churning cream
- Buttermilk, the liquid left over after producing butter from cream, often dried as *livestock* feed
- *Ghee*, clarified butter, by gentle heating of butter and removal of the solid matter
- *Smen*, a fermented, clarified butter used in Moroccan cooking
- *Anhydrous* milkfat (*clarified butter*)

◆ *Cheese*, produced by coagulating milk, separating from whey and letting it ripen, generally with *bacteria* and sometimes also with certain *molds*
- *Curds*, the soft, curdled part of milk (or skim milk) used to make cheese
- *Paneer*
- *Whey*, the liquid drained from curds and used for further processing or as a livestock feed

- *Cottage cheese*
- *Quark*
- *Cream cheese*, produced by the addition of cream to milk and then curdled to form a rich curd or cheese
- *Fromage frais*

♦ *Casein*
  - Caseinates, sodium or calcium salts of casein
  - Milk protein concentrates and isolates
  - Whey protein concentrates and isolates, reduced lactose whey
  - Hydrolysates, milk treated with proteolytic enzymes to alter functionality.
  - *Mineral* concentrates, byproduct of demineralizing whey

♦ *Yogurt*, milk fermented by *Streptococcus salivarius* ssp. *thermophilus* and *Lactobacillus delbrueckii ssp. bulgaricus* sometimes with additional bacteria, such

♦ as *Lactobacillus acidophilus*
  - *Ayran*
  - *Lassi*
  - *Leben*

♦ *Clabber*, milk naturally fermented to a yogurt-like state

♦ *Gelato*, slowly frozen milk and water, lesser fat than ice cream

♦ *Ice cream*, slowly frozen cream, milk, flavors and emulsifying additives
  - *Ice milk*, low-fat version of ice cream
  - *Frozen custard*
  - *Frozen yogurt*, yogurt with emulsifiers

♦ Other
  - *Viili*
  - *Kajmak*
  - *Filmjölk*
  - *Piimä*

- *Vla*
- *Dulce de leche*
- *Skyr*

DAIRY PRODUCTS

Name	Image	Origin	Description
Amasi			The common word for fermented milk that tastes like cottage cheese or plain yogurt. It is very popular in South Africa.
Ayran		Turkey	A Turkish beverage of yogurt mixed with cold water and sometimes salt that may be considered a variant of a drink popular throughout Central Asian, the Middle East, and South-eastern Europe. Ayran is found in the Balkans as well as Turkey and may be present in the North Caucasus, too.[2]
Baked milk			A variety of boiled milk that has been particularly popular in Russia, Ukraine and Belarus. It is made by simmering milk on low heat for eight hours or longer.
Basundi		India	An Indian dessert mostly in Bihar, Maharashtra, Gujarat and Karnataka. It is sweetened dense milk made by boiling milk on low heat until the milk is reduced by half.
Bhuna khoya		Khan garh, Pakistan	A type of khoa specially linked to city of Khan garh in Pakistan.
Blaand		Introduced to Scotland by Vikings	A fermented milk product made from whey. It is similar in alcohol content to wine.
Black Kashk		Central Asia	Prepared from yogurt, its production involves several processes.
Booza			An elastic, sticky, high level melt resistant ice cream which should delay melting in the hotter climates of the Arabic countries where it is most commonly found.
Buffalo curd			A traditional and nutritious dairy product prepared from buffalo milk and it is popular throughout south Asian countries such as India, Pakistan, Sri Lanka and Nepal.
Bulgarian yogurt		Bulgaria	A fermented milk product. In common with all dairy yogurt, Bulgarian yogurt is produced through the bacterial fermentation of milk, using a live culture of Lactobacillus bulgaricus and Streptococcus thermophilus.
Butter			Made by churning fresh or fermented cream or milk. It is generally used as a spread and a condiment, as well as in cooking, such as baking, sauce making, and pan frying. Butter

Contd...

Name	Image	Origin	Description
			consists of butterfat, milk proteins and water.
Butterfat			The fatty portion of milk. Milk and cream are often sold according to the amount of butterfat they contain.
Buttermilk			Refers to a number of dairy drinks. Originally, buttermilk was the liquid left behind after churning butter out of cream. This type of buttermilk is known as traditional buttermilk.
Buttermilk koldskål		Denmark	A sweet cold beverage or soup, made with buttermilk and other ingredients. Pictured is buttermilk koldskål with biscuits.
Buttermilk powder			Used in the production of ice cream as a source of solids, in processed sliced cheese to increase viscosity, as an emulsifier in chocolate products and in dry mixes such as pancake mix, to add dairy flavor and enhance food browning.
Cacik		Turkey	A Turkish dish of seasoned, diluted yogurt, eaten throughout the former Ottoman countries. In Greece a similar, much thicker yogurt dish is calledtzatziki and is also similar to tarator in Balkan cuisine.
Camel milk			Camel's milk has supported Bedouin, nomad and pastoral cultures since the domestication of camels' millennia ago. Herders may for periods survive solely on the milk when taking the camels on long distances to graze in desert and arid environments. Camel dairy farming is an alternative to cow milk in dry regions of the world.
Casein			The name for a family of related phosphoproteins (α S_1, βS_2, β, κ). These proteins are commonly found in mammalian milk, making up 80% of the proteins in cow milk and between 20% and 45% of the proteins in human milk.
Caudle			A British thickened and sweetened alcoholic hot drink, somewhat like eggnog. It was popular in the Middle Ages for its supposed medicinal properties.
Chaas			A buttermilk preparation from India. It is consumed all year round where it is usually taken along with meals. It contains raw milk, cream (malai) or yogurt which is blended manually in a pot with an instrument called madhani (whipper).
Chal			A Turkic (especially Turkmen and Kazakh) beverage of fermented camel milk, sparkling white with a sour flavor, popular in Central Asia—particularly in Kazakhstan and Turkmenistan. In the image, chal is pictured left, along with kumis on the right.

Contd...

Name	Image	Origin	Description
Chalap			A beverage common to Kyrgyzstan and Kazakhstan. It consists of yogurt, salt, and modernly, carbonated water.
Chass			The word used for buttermilk in Rajasthani and Gujarati. Chass is the traditional Gujarati beverage from Gujarat, India. It is similar to, but cheaper than, Lassi.
Cheese			A food derived from milk that is produced in a wide range of flavors, textures, and forms by coagulation of the milk protein casein. It comprises proteins and fat from milk, usually the milk of cows, buffalo, goats, or sheep.
Clabber			Produced by allowing unpasteurized milk to turn sour at a specific humidity and temperature. Over time, the milk thickens or curdles into a yogurt-like substance with a strong, sour flavor.
Clotted cream			A thick cream made by indirectly heating full-cream cow's milk using steam or a water bath and then leaving it in shallow pans to cool slowly. During this time, the cream content rises to the surface and forms 'clots' or 'clouts'. It forms an essential part of a cream tea.
Condensed milk			Milk from which water has been removed. It is most often found in the form of sweetened condensed milk, with sugar added.
Cottage cheese			A cheese curd product with a mild flavor. It is drained, but not pressed, so some whey remains and the individual curds remain loose.
Cream			Composed of the higher-butterfat layer skimmed from the top of milk before homogenization. In un-homogenized milk, the fat, which is less dense, will eventually rise to the top.
Cream cheese			A soft, mild-tasting cheese with a high fat content. Traditionally, it is made from unskimmed milk enriched with additional cream. Stabilizers such as carob bean gum and carrageenan are added.
Crème anglaise			A light pouring custard used as a dessert cream or sauce. It is a mix of sugar, egg yolks and hot milk, often flavored with vanilla.
Crème fraîche			A soured cream containing 30–45% butterfat and having a pH of around 4.5. It is soured with bacterial culture, but is less sour than

Contd..

Name	Image	Origin	Description
			U.S.-stylesour cream, and has a lower viscosity and a higher fat content.
Cuajada		Spain	A milk curd) cheese product. Traditionally it is made from ewe's milk, but now it is more often made industrially from cow's milk. It is popular in the north-eastern regions of Spain (Basque Country, Navarre, Castilla y León, La Rioja).
Curd			Curd is obtained by curdling (coagulating) milk with rennet or an edible acidic substance such as lemon juice or vinegar, and then draining off the liquid portion. The increased acidity causes the milk proteins (casein) to tangle into solid masses, or curds.
Curd snack			A type of sweet snack made from curd, popular in the Baltic states – Estonia, Latvia and Lithuania – as well as in Russia, Belarus, Ukraine and Kazakhstan.
Custard			A variety of culinary preparations based on a cooked mixture of milk or cream and egg yolk. Depending on how much egg or thickener is used, custard may vary in consistency from a thin pouring sauce to a thick pastry cream used to fill éclairs.
Dadiah		West Sumatra, Indonesia	A traditional fermented milk of West Sumatra made by pouring fresh raw unheated buffalo milk into a bamboo tube capped with a banana leaf, and allowing it to ferment spontaneously at room temperature for two days.
Daigo		Japan	A type of dairy product made in Japan during the 10th century.
Dondurma		Turkey	The name given to ice cream in Turkey. Dondurma typically includes the ingredients milk, sugar, salep, and mastic.
Donkey milk			The milk given by the domesticated ass or donkey. It has been used since Egyptian antiquity for both alimentary and cosmetic reasons.
Doogh			A yogurt-based beverage. Popular in Iran, Afghanistan, Azerbaijan, Armenia, Iraq, Syria and Turkey, it is sometimes carbonated. Outside of Iran and Afghanistan it is known by different names.
Evaporated milk			Also known as dehydrated milk, evaporated milk is a shelf-stable canned milk product with about 60% of the water removed from fresh milk. It differs from sweetened condensed milk, which contains added sugar.

Contd...

Name	Image	Origin	Description
Filled milk			Any milk, cream, or skim milk that has been reconstituted with fats, usually vegetable oils, from sources other than dairy cows.
Filmjölk		Scandinavia	A Nordic dairy product, similar to yogurt, but using different bacteria which give a different taste and texture.
Fromage frais		north of France and the south of Belgium	The name means "fresh cheese" in French (fromage blanc translates as "white cheese").
Fermented milk products			Also known as cultured dairy foods, cultured dairy products, or cultured milk products, fermented milk products are dairy foods that have been fermented with lactic acid bacteria such as Lactobacillus, Lactococcus, and Leuconostoc. Pictured is skyr.
Frozen custard			A cold dessert similar to ice cream, but made with eggs in addition to cream and sugar.
Frozen yogurt		United States	A frozen dessert made with yogurt and sometimes other dairy products. It varies from slightly to much more tart than ice cream, as well as being lower in fat (due to the use of milk instead of cream).
Galalith			A synthetic plastic material manufactured by the interaction of casein and formaldehyde
Gelato		Italy	The Italian word for ice cream, derived from the Latin word "gelâtus." (meaning frozen). Gelato is made with milk, cream, various sugars, and flavoring such as fresh fruit and nut purees.
Goat milk			Goats produce about 2% of the world's total annual milk supply. Some goats are bred specifically for milk.
Gombe		Sogn og Fjordane, Norway	A traditional dish from Sogn og Fjordane in Norway, it's prepared from curdled unpasteurized milk which is boiled down with sugar for several hours.
Gomme		Norway	A traditional Norwegian dish used for dinner or dessert, gomme is a sort of sweet cheese made of long-boiled milk, having a yellow or brown color. A white, porridge-like variant made of milk and oat grains or rice also exists.
Horse milk			Products collected from living horses include mare's milk, used by people with large horse herds, such as the Mongols, who let it ferment to produce kumis.
Ice cream			A frozen dessert usually made from dairy products, such as milk and cream and often combined with fruits or other ingredients and flavors.

Contd...

Name	Image	Origin	Description
Ice milk			A frozen dessert with less than 10 percent milkfat and the same sweetener content as ice cream.
Indian dairy products			A variety of dairy projects are indigenous to India and an important part of Indian cuisine. The majority of these products can be broadly classified into curdled products, like chhena, or non-curdled products, like khoa. Pictured is Paneer.
Infant formula			A manufactured food designed and marketed for feeding to babies and infants under 12 months of age, usually prepared for bottle-feeding or cup-feeding from powder (mixed with water) or liquid (with or without additional water).
Junket			A milk-based dessert, made with sweetened milk and rennet, the digestive enzyme which curdles milk.
Kashk			A large family of foods found in Armenian, Iranian, Lebanese, Palestinian, and Syrian cuisines. There are three main kinds of food with this name: foods based on curdled milk products like yogurt or cheese are within the realm of dairy products. Pictured are Kurdish women preparing Kashk.
Kaymak			A creamy dairy product, similar to clotted cream. It is made from the milk of water buffalos or of cows.
Kefir			A fermented milk drink prepared by inoculating cow, goat, or sheep milk with kefir grains.
Khoa			A milk food widely used in Indian and Pakistani cuisine, made of either dried whole milk or milk thickened by heating in an open iron pan.
Kulfi			A popular frozen dairy dessert from the Indian Subcontinent. It is often described as "traditional Indian Subcontinent ice cream".
Kumis			A fermented dairy product traditionally made from mare's milk. The drink remains important to the peoples of the Central Asian steppes, of Huno-Bulgar, Turkic and Mongol origin: Bashkirs, Kalmyks, Kazakhs, Kyrgyz, Mongols, Uyghurs, and Yakuts.
Lassi		India's Punjab region	A popular, traditional, yogurt-based drink consisting of a blend of yogurt, water, spices, and sometimes, fruit.
Leben			A fermented milk product commonly available in the Arab world

Contd...

Name	Image	Origin	Description
Malai			An Indian term for clotted cream or Devonshire cream. It is made by heating non-homogenized whole milk to about 80°C (180°F) for about one hour and then allowing to cool.
Matzoon		Armenia	A fermented milk product of Armenian origin made from cow's milk (mostly), goat's milk, sheep's milk, or a mix of them and a culture from previous productions.
Milk			A white liquid produced by the mammary glands of mammals. It is the primary source of nutrition for young mammals before they are able to digest other types of food.
Milk skin			A sticky film of protein that forms on top of milk and milk-containing liquids (such as hot chocolate and some soups). It is caused by the denaturation of proteins such as casein. In Japan, a dairy product called So was made from layers of milk skin during the 7th-10th centuries.
Mitha Dahi			A fermented sweet dahi or sweet yogurt. This type of yogurt is common in the states of West Bengal and Odisha in India, and in Bangladesh.
Moose milk			Pictured is a milkmaid collecting moose milk at Kostroma Moose Farm in Kostroma Oblast, Russia.
Mursik		Kenya	A basic element of the cuisine of the Kalenjin people of Kenya. Made from curdled dairy products cooked in a specially made gourd container, it is commonly served at dinner.
Pomazánkové máslo			A traditional Czech and Slovak dairy product, it is a spread made from base ingredients of sour cream, milk powder and buttermilk powder.
Powdered milk			A manufactured dairy product made by evaporating milk to dryness. One purpose of drying milk is to preserve it; milk powder has a far longer shelf life than liquid milk and does not need to be refrigerated, due to its low moisture content.
Processed cheese			A food product made from normal cheese and sometimes other unfermented dairy ingredients, plus emulsifiers, extra salt, food colorings, or whey. Many flavors, colors, and textures of processed cheese exist.
Pytia			Curdled milk obtained from an animal's stomach, containing (and used as) rennet.
Qimiq			Consists of 99% light cream and 1% gelatin; it was invented in 1995 and is patented by Hama Foodservice GmbH.

Contd...

Name	Image	Origin	Description
Quark			A fresh dairy product made by warming soured milk until the desired degree of denaturation of milk proteins is met, and then strained. It is soft, white and unaged, similar to some types of fromage frais.
Reindeer milk			Reindeer have been herded for centuries by several Arctic and Subarctic people including the Sami and the Nenets. They are raised for their meat, hides, and antlers and, to a lesser extent, for milk and transportation.
Ryazhenka		Ukraine	Fermented baked milk
Semifreddo			A class of semi-frozen desserts, typically ice-cream cakes, semi-frozen custards, and certain fruit tarts. It has the texture of frozen moussebecause it is usually produced by uniting two equal parts of ice cream and whipped cream.
Sergem		Tibet	A Tibetan food made from milk once the butter from the milk is extracted. It is then put in a vessel and heated and when it is about to boil, sour liquid call "chakeu" is add and this leads to the separation of sergem from that milk.
Sheep milk			Also known as ewe's milk, it's the milk of domestic sheep. Though not widely drunk in any modern culture, sheep's milk is commonly used to make cultured dairy products.
Shrikhand		India	An Indian sweet dish made of strained yogurt. It is one of the main desserts in Maharashtrian cuisine and Gujarati cuisine.
Skorup			Kajmak that is matured in dried animal skin sacks is called skorup.
Skyr		Iceland	An Icelandic cultured dairy product, similar to strained yogurt. It has been a part of Icelandic cuisine for over a thousand years.
Smântână		Romania	A Romanian dairy product that is produced by souring heavy cream.
Smetana			A range of sour creams from Central and Eastern Europe. It is a dairy product produced by souring heavy cream. Pictured is a bowl of borschtsoup topped with smetana.
So		Japan	A type of dairy product that was made in Japan between 7th and 10th centuries.
Soft serve		United States	A type of ice cream that is softer than regular ice cream, as a result of air being introduced during freezing. Soft serve ice cream has been sold commercially since the late 1930s.

Contd...

Name	Image	Origin	Description
Sour cream			Obtained by fermenting a regular cream with certain kinds of lactic acid bacteria. The bacterial culture, which is introduced either deliberately or naturally, sours and thickens the cream.
Soured milk			Produced from the acidification of milk. It is not the same as spoiled milk that has soured naturally and which may contain toxins. Acidification, which gives the milk a tart taste, is achieved either through the addition of an acid, such as lemon juice or vinegar, or through bacterialfermentation.
Spaghettieis		Mannheim, Germany	A German ice cream made to look like a plate of spaghetti. It was created by Dario Fontanella in the late 1960s in Mannheim, Germany.
Stewler			A fermented milk product that is popular in Russia and Ukraine. Similar to Ryazhenka, it is made by adding sour cream to baked milk.
Strained yogurt			Yogurt which has been strained in a cloth or paper bag or filter to remove the whey, giving a consistency between that of yogurt and cheese, while preserving yogurt's distinctive sour taste. Pictured is strained yogurt with olive oil.
Súrmjólk		Iceland	A cultured milk product or a type of yogurt. It is made from either whole or semi-skimmed milk and various flavorings are sometimes added.
Uunijuusto		Finland	A dish made from cow's colostrum, the first milk of a calved cow, which has salt added and is then baked in an oven.
Vaccenic acid			An omega-7 fatty acid. It is a naturally occurring trans-fatty acid found in the fat of ruminants and in dairy products such as milk, butter, and yogurt.
Viili			A yogurt-like mesophilic fermented milk that originated in the Nordic countries. It has a ropey, gelatinous consistency and a pleasantly mild taste resulting from lactic acid.
Vla		Netherlands	A type of custard (known in the United States as cornstarch pudding).
Whey			The liquid remaining after milk has been curdled and strained. It is a by-product of the manufacture of cheese or casein and has several commercial uses.
Whey protein			A mixture of globular proteins isolated from whey, the liquid material created as a by-product of cheese production.

Contd...

Name	Image	Origin	Description
Whipped cream			Cream that has been beaten by a mixer, whisk, or fork until it is light and fluffy. Whipped cream is often sweetened and sometimes flavored withvanilla, and is often called Chantilly cream or crème Chantilly.
Yak butter			A staple food item and trade item for herding communities in south Central Asia and the Tibetan Plateau. Many different political entities have communities of herders who produce and consume yak's dairy products including cheese and butter – for example, China, India, Mongolia, Nepal, and Tibet.
Yak milk			Domesticated yaks have been kept for thousands of years, primarily for their milk, fibre and meat, and as beasts of burden.
Yakult		Created by Japanes scientist Minoru Shirota	A probiotic dairy product made by fermenting a mixture of skimmed milk with a special strain of the bacterium Lactobacillus casei Shirota.
Ymer		Denmark	A Danish soured milk product which has been known since 1930. It is made by fermenting whole milk with the bacterial culture Lactococcus lactis.
Yogurt			A fermented milk product (soy milk, nut milks such as almond milk, and coconut milk can also be used) produced by bacterialfermentation of milk. The bacteria used to make yogurt are known as "yogurt cultures".

POULTRY

General Information

India has made considerable progress in broiler production in the last two decades. High quality chicks, equipments, vaccines and medicines are available. With an annual output of 41.06 billion eggs and 1000 million broilers, India ranks fourth largest producer of eggs and fifth largest producer of poultry broiler in the world. The broiler production has also sky rocketed at an annual growth rate of about 15 percent at present. Broiler farming has been given considerable importance in the national policy and has a good scope for further development in the years to come.

Advantages of Chicken farming

- Initial investment is a little lower than layer farming.

- Rearing period is 6-7 weeks only.

- More number of flocks can be taken in the same shed.

- Broilers have high feed conversion efficiency i.e. least amount of feed is required for unit body weight gain in comparison to other livestock.

- Faster return from the investment.

- Demand for poultry meat is more compared to sheep/Goat meat.

Indigenous Breeds

- The common control hen, the desi, is as a rule the best mother for hatching. She is a good forager. Some of the Indian flows resemble the Leghorn in size and shape, but have poor laying qualities. They are Found in various colours. one variety found in India resembles the sussex or Plymouth Rock in shape but is smaller. These birds lay family well and are more common in the eastern parts of the country.

- The Indian birds are mostly non-descripts, and are of very little value as layers. They have several local breed names such as Tenis, Naked Neck Punjab, Brown, Ghagus, Lolab, Kashmir Faberella, Tilri, Busra Telllicherry, Danki, Nicorai and Kalahasti. There are only 4 pure breeds Karaknath and the Busra. The last occurs in western India. A large number of flows of different size, shapes and colours, and for the most part resembling the jungle fowls, are found all over India. They vary in appearance according to the locality in which they have been bred. These with Chittagong, Aseel, Langshan or Brahma blood in them are bigger in size and better in meat quality than the common flows.

Aseel

Aseel is noted for its pugnacity, high stamina, majestic gait and dogged fighting qualities. The best specimens of the breed, although rare, and encountered in parts of Andhra Pradesh, Uttar Pradesh and Rsjasthan. The most popular varieties are peela (golden red), yarkin (black and red), Nurie 89(white), kagar (black), chitta (black and white silver), Teekar (brown) and Reza (light red). Although poor in productivity, the birds of this breed are well-known for their meat qualities. Broodiness in most common and the hen is a good sitter and efficient mother. They possess pea combs which are small but firmly set on head. Wattles and ear lobes are bright red, and the beak is hart. The face is long and slender, and not covered with feather. The eyes are compact, well set and present bold looks. The neck is long, uniformly thick but no fleshy. The body is round and short with broad breast straight back and close - set strong tail root. The general feathering is close, scanty and almost absent on the Brest. The plumage has practically no fluff and the feathers are tough. The tail is small and drooping. The legs are strong, straight, and set well apart. Standard weight (kg): Cocks, 4 to 5; hen 3 to 4; cockerrels, 3.5 to 4.5; pullets, 2.5 to 3.5.

Karaknath

The original name of the breed seems to be Kalamasi, meaning a fowl with black flesh. However, it is popularly known as Karaknath. The eggs are light brown. The day-old chicks are bluish to black with irregular dark stripes over the back. The adult plumage varies from silver and gold-spangled to bluish-black without any spangling. The skin, beak, shanks, toes and soles of feet are slatelike in colour. The comb, wattles and tongue are purple. Most of the internal organs show intense black colouration which is pronounced in trachea, thoracic and abdominal air-sacs, gonads and at the base of the heart and mesentery. Varying degrees of block colouration are also seen in the skeletal muscles, tendons, nerves, meanings, brain etc. The blood is darker than normal blood. The black pigment has been due to deposition of melanin; The flesh although repulsive to look at, is delicious. A medium layer lays about 80 eggs per year. The bird is resistant to diseases in its natural habitat in free range but is more susceptible to Mareks disease under intensive rearing conditions.

Other commercial breeds of broiler chicken in India

Breed	First egg	50% Production	Peak production	Livability	Egg production peak	Feed efficiency	Egg weight	Net egg production (72 weeks)
ILI-80	17-18 weeks	150 days	26-28 weeks	Grower (96%) Layer (94%)	92%	2.1	54 g	280 eggs
Golden-92	18-19 weeks	155 days	27-29 weeks	Grower (96%) Layer (94%)	90%	2.2	54 g	265 eggs
Priya	17-18 weeks	150 days	26-28 weeks	Grower (96%) Layer (94%)	92%	2.1	57 g	290 eggs
Sonali	18-19 weeks	155 days	27-29 weeks	Grower (96%) Layer (94%)	90%	2.2	54 g	275 eggs
Devendra	18-19 weeks	155 days	27-29 weeks	Grower (97%) Layer (94%)	90%	2.5	50 g	200 eggs

Commercial available meat-type chicken in India

Breed	Weight at six weeks (g)	Weight at seven weeks(g)	Food conversion ratio	Livability (%)
B-77	1300	1600	2.3	98-99
CARIBRO-91	1650	2100	1.94-2.2	97-98
CARIBRO Multicoloured	1600	2000	1.9-2.1	97-98
CARIBRO Naked necked	1650	2000	1.9-2.0	97-98
Varna	1500	1800	2.1-2.25	97

POULTRY FARMING BUSINESS

Poultry farming is the raising of domesticated birds such as chickens, turkeys, ducks, and geese, for the purpose of farming meat or eggs for food. Poultry are farmed in great numbers with chickens being the most numerous. More than 50 billion chickens are raised annually as a source of food, for both their meat and their eggs. Chickens raised for eggs are usually called layers while chickens raised for meat are often called broilers. In total, the UK alone consumes over 29 million eggs per day. In the US, the national organization overseeing poultry production is the Food and Drug Administration (FDA). In the UK, the national organization is the Department for Environment, Food and Rural Affairs (Defra).

Intensive and alternative poultry farming

According to the World watch Institute, 74 percent of the world's poultry meat, and 68 percent of eggs are produced in ways that are described as 'intensive'. One alternative to intensive poultry farming is free-range farming using lower stocking densities. all seasons at a lower cost than free-range production. Poultry producers routinely use nationally approved medications, such as antibiotics, in feed or drinking water, to treat disease or to prevent disease outbreaks. Some FDA-approved medications are also approved for improved feed utilization.

Egg-laying chickens' husbandry systems

Commercial hens usually begin laying eggs at 16–20 weeks of age, although production gradually declines soon after from approximately 25 weeks of age. This means that in many countries, by approximately 72 weeks of age, flocks are considered economically unviable and are slaughtered after approximately 12 months of egg production, although chickens will naturally live for 6 or more years. In some countries, hens are force molted to re-invigorate egg-laying.

Environmental conditions are often automatically controlled in egg-laying systems. For example, the duration of the light phase is initially increased to prompt the beginning of egg-laying at 16–20 weeks of age and then mimics summer day length which stimulates the hens to continue laying all year round; normally, egg production occurs only in the warmer months. Some commercial breeds of hen can produce over 300 eggs a year. Critics argue that year-round egg production stresses the birds more than normal seasonal production.

Free-range

Free-range poultry farming allows chickens to roam freely for a period of the day, although they are usually confined in sheds at night to protect them from predators or kept indoors if the weather is particularly bad. In the UK, the Department for Environment, Food and Rural Affairs (Defra) states that a free-range chicken must have day-time access to open-air runs during at least half of its life. Unlike in the United States, this definition also applies to free-range egg laying hens. The European Union regulates marketing standards for egg farming which specifies a minimum condition for free-range eggs that "hens have continuous daytime access to open-air runs, except in the case of temporary restrictions imposed by veterinary authorities". The RSPCA "Welfare standards for laying hens and pullets" indicates that the stocking rate must not exceed 1,000 birds per hectare (10 m² per hen) of range available and a minimum area of overhead shade/shelter of 8 m² per 1,000 hens must be provided.

Free-range farming of egg-laying hens is increasing its share of the market. Defra figures indicate that 45% of eggs produced in the UK throughout 2010 were free-range, 5% were produced in barn systems and 50% from cages. This compares with 41% being free-range in 2009.

Suitable land requires adequate drainage to minimize worms and coccidial oocysts, suitable protection from prevailing winds, good ventilation, access and protection from predators. Excess heat, cold or damp can have a harmful effect on the animals and their productivity. Free-range farmers have less control than farmers using cages in what food their chickens eat, which can lead to unreliable productivity, though supplementary feeding reduces this uncertainty. In some farms, the manure from free-range poultry can be used to benefit crops.

The benefits of free-range poultry farming for laying hens include opportunities for natural behaviours such as pecking, scratching, foraging and exercise outdoors. Both intensive and free-range farming have animal welfare concerns. Cannibalism, feather pecking and vent pecking can be common with some farmers using beak trimming as a preventative measure, although reducing stocking rates would eliminate these problems. Diseases can be common and the animals are vulnerable to predators. Barn systems have been found to have the worst bird welfare. In South-East Asia, a lack of disease control in free range farming has been associated with outbreaks of Avian influenza.

Organic

In organic egg-laying systems, chickens are also free-range. Organic systems are based upon restrictions on the routine use of synthetic yolk colourants, in-feed or in-water medications, other food additives and synthetic amino acids, and a lower stocking density and smaller group sizes. The Soil Association standards used to certify organic flocks in the UK indicate a maximum outdoors stocking density of 1,000 birds per hectare and a maximum of 2,000 hens in each poultry house. In the UK, organic laying hens are not routinely beak-trimmed.

Yarding

While often confused with free-range farming, yarding is actually a separate method of poultry culture by which chickens and cows are raised together. The distinction is that free-range poultry are either totally unfenced, or the fence is so distant that it has little influence on their freedom of movement. Yarding is common technique used by small farms in the Northeastern US. The birds are released daily from hutches or coops. The hens usually lay eggs either on the floor of the coop or in baskets if provided by the farmer. This husbandry technique can be complicated if used with roosters, mostly because of aggressive behaviour.

Battery cage

The majority of hens in many countries are reared in battery cages, although the European Union Council Directive 1999/74/EC has banned the conventional battery cage in EU states from January 2012. These are small cages, usually made of metal in modern systems, housing 3 to 8 hens. The walls are made of either solid metal or mesh, and the floor is sloped wire mesh to allow the faces to drop through and eggs to roll onto an egg-collecting conveyor belt. Water is usually provided by overhead nipple systems, and food in a trough along the front of the cage replenished at regular intervals by a mechanical chain. The cages are arranged in long rows as multiple tiers, often with cages back-to-back (hence the term 'battery cage'). Within a single shed, there may be several floors containing battery cages meaning that a single shed may contain many tens of thousands of hens. Light intensity is often kept low (e.g. 10 lux) to reduce feather pecking and vent pecking. Benefits of battery cages include easier care for the birds, floor eggs which are expensive to collect are eliminated, eggs are cleaner, capture at the end of lay is expedited, generally less feed is required to produce eggs, broodiness is eliminated, more hens may be housed in a given house floor space, internal parasites are more easily treated, and labor requirements are generally much reduced.

Furnished cage

In farms using cages for egg production, there are more birds per unit area; this allows for greater productivity and lower food costs. Floor space ranges upwards from 300 cm^2 per hen. EU standards in 2003 called for at least 550 cm^2 per hen. In the US, the current recommendation by the United Egg Producers is 67 to 86 in^2 (430 to 560 cm^2) per bird. The space available to battery hens has often been described as less than the size of a piece of A4 paper. Animal welfare scientists have been critical of battery cages because they do not provide hens with sufficient space to stand, walk, flap their wings, perch, or make a nest, and it is widely considered that hens suffer through boredom and frustration through being unable to perform these behaviours. This can lead to a wide range of abnormal behaviours, some of which are injurious to the hens or their cagemates.

In 1999, the European Union Council Directive 1999/74/EC banned conventional battery cages for laying hens throughout the European Union from January 1, 2012; they were banned previously in other countries including Switzerland. In response to these bans, development of prototype commercial furnished cage systems began in the 1980s. Furnished cages, sometimes called 'enriched' or 'modified' cages, are cages for egg laying hens which have been designed to overcome some of the welfare concerns of battery cages whilst retaining their economic and husbandry advantages, and also provide some of the welfare advantages of non-cage systems. Many design features of furnished cages have been incorporated because research in animal welfare science has shown them to be of benefit to the hens. In the UK, the Defra "Code for the Welfare of Laying Hens" states furnished cages should provide at least 750 cm^2 of cage area per hen, 600 cm^2 of which should be usable; the height of the cage other than that above the usable area should be at least 20 cm at every point and no cage should have a total area that is less than 2000 cm^2. In addition, furnished cages should provide a nest, litter such that pecking and scratching are possible, appropriate perches allowing at least 15 cm per hen, a claw-shortening device, and a feed trough which may be used without restriction providing 12 cm per hen.

Modern egg laying breeds often suffer from osteoporosis which results in the chicken's skeletal system being weakened. During egg production, large amounts of calcium are transferred from bones to create egg-shell. Although dietary calcium levels are adequate, absorption of dietary calcium is not always sufficient, given the intensity of production, to fully replenish bone calcium. This can lead to increases in bone breakages, particularly when the hens are being removed from cages at the end of laying.

Meat Producing Chickens Husbandry Systems

Indoor broilers

Meat chickens, commonly called broilers, are floor-raised on litter such as wood shavings or rice hulls, indoors in climate-controlled housing. Under modern farming methods, meat chickens reared indoors reach slaughter weight at 5 to 6 weeks of age.

Broilers are not raised in cages. They are raised in large, open structures known as grow out houses. These houses are equipped with mechanical systems to deliver feed and water to the birds. They have ventilation systems and heaters that function as needed. The floor of the house is covered with bedding material consisting of wood chips, rice hulls, or peanut shells. Because dry bedding helps maintain flock health, most grow out houses have enclosed watering systems ("nipple drinkers") which reduce spillage.

Keeping birds inside a house protects them from predators such as hawks and foxes. Some houses are equipped with curtain walls, which can be rolled up in good weather to admit natural light and fresh air. Most grow out houses built in recent years feature "tunnel ventilation," in which a bank of fans draws fresh air through the house.

Traditionally, a flock of broilers consist of about 20,000 birds in a grow out house that measures 400 feet long and 40 feet wide, thus providing about eight-tenths of a square foot per bird. The Council for Agricultural Science and Technology (CAST) states that the minimum space is one-half square foot per bird. More modern houses are often larger and contain more birds, but the floor space allotment still meets the needs of the birds. Because broilers are relatively young and have not reached sexual maturity, they exhibit very little aggressive conduct. Chicken feed consists primarily of corn and soybean meal with the addition of essential vitamins and minerals. No hormones or steroids are allowed in raising chickens.

Issues with indoor husbandry

In intensive broiler sheds, the air can become highly polluted with ammonia from the droppings. This can damage the chickens' eyes and respiratory systems and can cause painful burns on their legs (called *hock burns*) and feet. Broilers bred for fast growth have a high rate of leg deformities because the large breast muscles cause's distortions of the developing legs and pelvis, and the birds cannot support their increased body weight. Because they cannot move easily, the chickens are not able to adjust their environment to avoid heat, cold or dirt as they would in natural conditions. The added weight and overcrowding also puts a strain on their hearts and lungs and Ascites can develop. In the UK, up to 19 million broilers die in their sheds from heart failure each year.

Indoor with higher welfare

Chickens are kept indoors but with more space (around 12 to 14 birds per square meter). They have a richer environment for example with natural light or straw bales that encourage foraging and perching. The chickens grow more slowly and live for up to two weeks longer than intensively farmed birds. The benefits of higher welfare indoor systems are the reduced growth rate, less crowding and more opportunities for natural behaviour.

Free-range broilers

Free-range broilers are reared under similar conditions to free-range egg laying hens. The breeds grow more slowly than those used for indoor rearing and usually reach slaughter weight at approximately 8 weeks of age. In the EU, each chicken must have one square meter of outdoor space. The benefits of free-range poultry farming include opportunities for natural behaviours such as pecking, scratching, foraging and exercise outdoors. Because they grow slower and have opportunities for exercise, free-range broilers often have better leg and heart health.

Organic broilers

Organic broiler chickens are reared under similar conditions to free-range broilers but with restrictions on the routine use of in-feed or in-water medications, other food additives and synthetic amino acids. The breeds used are slower growing, more traditional breeds and typically reach slaughter weight at around 12 weeks of age. They have a larger space allowance outside (at least 2 square meters and sometimes up to 10 square meters per bird). The Soil Association standards indicate a maximum outdoors stocking density of 2,500 birds per hectare and a maximum of 1,000 broilers per poultry house.

ISSUES WITH POULTRY FARMING

Humane treatment

Animal welfare groups have frequently criticized the poultry industry for engaging in practices which they believe to be inhumane. Many animal rights advocates object to killing chickens for food, the "factory farm conditions" under which they are raised, methods of transport, and slaughter. Compassion Over Killing and other groups have repeatedly conducted undercover investigations at chicken farms and slaughterhouses which they allege confirm their claims of cruelty. Conditions in chicken farms may be unsanitary, allowing the proliferation of diseases such as salmonella and E. coli. Chickens may be raised in very low light intensities, sometimes total darkness, to reduce injurious pecking. Concerns have been raised that companies growing single varieties of birds for eggs or meat are increasing their susceptibility to disease. Rough handling, crowded transport during various weather conditions and the failure of existing stunning systems to render the birds unconscious before slaughter; have also been cited as welfare concerns. A common practice among hatcheries for egg-laying hens is the culling of newly hatched male chicks since they don't lay eggs and do not grow fast enough to be profitable for meat.

Beak trimming

Laying hens are routinely beak-trimmed at 1 day of age to reduce the damaging effects of aggression, feather pecking and cannibalism. Scientific studies (see below) have shown that beak trimming is likely to cause both acute and chronic pain. The beak is a complex, functional organ with an extensive nervous supply including nociceptors that sense pain and noxious stimuli. These would almost certainly be stimulated during beak trimming, indicating strongly that acute pain would be experienced. Behavioral evidence of pain after beak trimming in layer hen chicks has been based on the observed reduction in pecking behavior, reduced activity and social behavior, and increased sleep duration. Severe beak trimming, or beak trimming birds at an older age, may cause chronic pain. Following beak trimming of older or adult hens, the nociceptors in the beak stump show abnormal patterns of neural discharge, which indicate acute pain. Neuromas, tangled masses of swollen regenerating axon sprouts, are found in the healed stumps of birds' beak trimmed at 5 weeks of age or older and in severely beak trimmed birds. Neuromas have been associated with phantom pain in human amputees and have therefore been linked to chronic pain in beak trimmed birds. If beak trimming is severe because of improper procedure or done in older birds, the neuromas will persist which suggests that beak trimmed older birds experience chronic pain, although this has been debated.

Beak-trimmed chicks will initially peck less than non-trimmed chickens, which animal behavioralist Temple Grandin attributes to guarding against pain. The animal rights activist, Peter Singer, claims this procedure is bad because beaks are sensitive, and the usual practice of trimming them without an aesthesia is considered inhumane by some. Some within the chicken industry claim that beak-trimming is not painful whereas others argue that the procedure causes chronic pain and discomfort, and decreases the ability to eat or drink.

Antibiotics

Antibiotics have been used on poultry in large quantities since the 1940s, when it was found that the byproducts of antibiotic production, fed because the antibiotic-producing mold had a high level of vitamin B_{12} after the antibiotics were removed, produced higher growth than could be accounted for by the vitamin B_{12} alone. Eventually it was discovered that the trace amounts of antibiotics remaining in the byproducts accounted for this growth. The mechanism is apparently the adjustment of intestinal flora, favoring "good" bacteria while suppressing "bad" bacteria that provoke inflammation of the gut mucosa. So, the goal of antibiotics as a growth promoter is the same as for probiotics. Because the antibiotics used are not absorbed by the gut, they do not put antibiotics into the meat or eggs.

Antibiotics are used routinely in poultry for this reason, and also to prevent and treat disease. Many contend that this puts humans at risk as bacterial strains develop stronger and stronger resistances. Critics point out that, after six decades of heavy agricultural use of antibiotics, opponents of antibiotics must still make arguments about theoretical risks, since actual examples are hard to come by. A proposed bill in the United States Congress would make the use of antibiotics in animal feed legal only for therapeutic (rather than preventative) use, but it has not been passed. However, this may present the risk of slaughtered chickens harboring pathogenic bacteria and passing them on to humans that consume them. In October 2000, the U.S. Food and Drug Administration (FDA) discovered that two antibiotics were no longer effective in treating diseases found in factory-farmed chickens; one antibiotic was swiftly pulled from the market, but the other, Baytril, was not. Bayer, the company which produced it, contested the claim and as a result, Baytril remained in use until July 2005. To prevent any residues of antibiotics in chicken meat, any given antibiotics are required to have a "withdrawal" period before they can be slaughtered. Samples of poultry at slaughter are randomly tested by the FSIS, and show a very low percentage of residue violations.

Arsenic

Poultry feed can also include *roxarsone* or *nitarsone, arsenical antimicrobial* drugs that also promote growth. *Roxarsone* was used as a *broiler* starter by about 70% of the broiler growers in between 1995 to 2000. The drugs have generated controversy because it contains *arsenic*, which is highly toxic to humans. This arsenic could be transmitted through run-off from the poultry yards. A 2004 study by the U.S. magazine Consumer Reports reported "no detectable arsenic in our samples of muscle" but found "A few of our chicken-liver samples has an amount that according to EPA standards could cause neurological problems in a child who ate 2 ounces of cooked *liver* per week or in an adult who ate 5.5 ounces per week." The U.S. Food and Drug Administration (FDA), however, is the organization responsible for the regulation of foods in America, and all samples tested were "far less than the amount allowed in a food product.

Growth hormones

Hormone use in poultry production is illegal in the United States. Similarly, no chicken meat for sale in Australia is fed hormones. Several scientific studies have documented the fact that chickens grow rapidly because they are bred to do so, not because of growth hormones. A small producer of natural and organic chickens confirmed this assumption: Using hormones to boost egg production was a brief fad in the Forties, but was abandoned because it didn't work. Using hormones to produce soft-meated roasters lasted into the Fifties, but the improved growth rates of normal, untreated broilers made the practice irrelevant—the broilers got as big as anyone wanted without chemicals. The only hormone that was ever used in any quantity on poultry (DES) was banned in 1959, and everyone but a few die-hard farmers had given up hormones by then, anyway. Hormones are now illegal in poultry and eggs.

E. coli

According to Consumer Reports, "1.1 million or more Americans are sickened each year by undercooked, tainted chicken." A USDA study discovered E. coli (Biotype I) in 99% of supermarket chicken, the result of chicken butchering not being a sterile process. However, the same study also shows that the strain of E. coli found was always a non-lethal form, and no chicken had any of the pathenogenic O157:H7 serotype. Many of these chickens, furthermore, had relatively low levels of contamination. Feces tend to leak from the carcass until the evisceration stage, and the evisceration stage itself gives an opportunity for the interior of the carcass to receive intestinal bacteria. (So do the skin of the carcass, but the skin presents a better barrier to bacteria and reaches higher temperatures during cooking). Before 1950, this was contained largely by not eviscerating the carcass at the time of butchering, deferring this until the time of retail sale or in the home. This gave the intestinal bacteria less opportunity to colonize the edible meat. The development of the "ready-to-cook broiler" in the 1950s added convenience while introducing risk, under the assumption that end-to-end refrigeration and thorough cooking would provide adequate protection. E. coli can be killed by proper cooking times, but there is still some risk associated with it, and its near-ubiquity in commercially farmed chicken is troubling to some. Irradiation has been proposed as a means of sterilizing chicken meat after butchering.

Avian influenza

There is also a risk that crowded conditions in chicken farms will allow avian influenza (bird flu) to spread quickly. A United Nations press release states: "Governments, local authorities and international agencies need to take a greatly increased role in combating the role of factory-farming, commerce in live poultry, and wildlife markets which provide ideal conditions for the virus to spread and mutate into a more dangerous form.

Efficiency

Farming of chickens on an industrial scale relies largely on high protein feeds derived from soybeans; in the European Union the soybean dominates the protein supply for animal feed, and the poultry industry is the largest consumer of such feed. Two kilograms of grain must be fed to poultry to produce 1 kg of weight gain, much less than that required for pork or beef. However, for every gram of protein consumed, chickens yield only 0.33 g of edible protein.

Economic factors

Changes in commodity prices for poultry feed have a direct effect on the cost of doing business in the poultry industry. For instance, a significant rise in the price of corn in the United States can put significant economic pressure on large industrial chicken farming operations.

World chicken population

The Food and Agriculture Organization of the United Nations estimated that in 2002 there were nearly sixteen billion chickens in the world, counting a total population of 15,853,900,000. The figures from the Global Livestock Production and Health Atlas for 2004 were as follows:

Global Livestock Production and Health Atlas for 2004

Sr. No.	Country	Chicken population (birds)
1.	China	3,860,000,000
2.	United States	1,970,000,000
3.	Indonesia	1,200,000,000
4.	Brazil	1,100,000,000
5.	India	648,830,000
6.	Mexico	540,000,000
7.	Russia	340,000,000
8.	Japan	286,000,000
9.	Iran	280,000,000
10.	Turkey	250,000,000
11.	Bangladesh	172,630,000
12.	Nigeria	143,500,000

The Different Pig Breeds

Breed	Origin	Height	Weight	Color
Aksai Black Pied	Kazakhstan	---	240-320 kg (530-710 lb)	Black & White
American Yorkshire	USA	---	---	White
Angeln Saddleback	Germany	---	300-350 kg (660-770 lb)	Black & White
Appalachian English	---	---	---	---
Arapawa Island	New Zealand	---	100-180 kg (220-400 lb)	Sandy/Tan
Auckland Island Pig	New Zealand	---	---	Black, Black & Tan
Australian Yorkshire	Australia	---	---	---
Babi Kampung	---	---	---	---
Ba Xuyen	Vietnam	---	100 kg (220 lb)	Black & White
Bantu	---	---	---	---
Basque	France	---	---	---
Bazna	Romania	---	---	Black & White
Beijing Black	China	---	---	Black
Belarus Black Pied	Belarus	---	---	Black & White
Belgian Landrace	Belgium	---	---	White
Bengali Brown Shannaj	---	---	---	---
Bentheim Black Pied	Germany	70-75 cm (28-30 in)	180-250 kg (400-550 lb)	Black & White
Berkshire	United Kingdom	---	---	Black with some White markings

Contd...

Breed	Origin	Height	Weight	Color
Bisaro	Portugal	---	---	Black & White
Black Slavonian	---	---	---	Black
Black Canarian Pig	---	---	---	Black
Breitovo	Russia	---	---	---
British Landrace	United Kingdom	---	---	White
British Lop	United Kingdom	---	---	White
British Saddleback	United Kingdom	---	---	Black & White
Bulgarian White	Bulgaria	---	---	White
Cantonese	---	---	---	---
Chato Murciano	Spain	---	180 kg (400 lb)	Black or White or Black & White
Chester White	USA	---	---	White
Choctaw Hog	USA	---	60 kg (130 lb)	Black
Creole Pig	Haiti	---	---	---
Cumberland Pig	United Kingdom	---	---	White
Czech Improved White	---	---	---	White
Danish Landrace	Denmark	---	---	White
Danish Protest Pig	Denmark	92 cm (36 in)	350 kg (770 lb)	Red & White
Dermantsi Pied	---	---	---	---
Li Yan Pig	---	Singapore Boon Lay	---	---
Duroc	USA	---	---	Red or Black
Dutch Landrace pig	Netherlands	---	---	---
East Balkan pig	---	---	---	---
Essex	United Kingdom	---	---	Black & White
Estonian Bacon	Estonia	---	---	---
Fengjing pig	China	---	---	---
Finnish Landrance	Finland	---	---	---
Forest Mountain	Armenia	---	165-260 kg (360-570 lb)	White
French Landrace	France	---	---	White
Gascon	France	---	---	Black
German Landrace	Germany	---	---	---
Gloucestershire Old Spot	United Kingdom	---	220-280 kg (490-620 lb)	White with Black spots
Grice	United Kingdom	---	---	---
Guinea Hog	USA	---	200 kg (440 lb)	Black
Hampshire	United Kingdom	---	---	Black & White
Hante	---	---	---	---

Contd...

Breed	Origin	Height	Weight	Color
Hereford	USA	---	200-370 kg (440-820 lb)	Red & White
Hezuo	---	---	---	---
Iberian	Iberian Peninsula	---	---	Black, Red
Italian Landrace	Italy	---	---	White
Japanese Landrace	Japan	---	---	---
Jeju Black Pig	Jeju Island, South Korea	---	---	Black
Jinhua	---	---	---	---
Kakhetian	Georgia	---	---	---
Kele	---	---	---	---
Kemerovo	Russia	---	---	---
Korean Native Pig	Korea	---	---	Black
Krskopolje	---	---	---	---
Kunekune	New Zealand	60 cm (24 in)	68-100 kg (150-220 lb)	Red, Black & White, Cream, Gold-tip, Black, Brown & Tri-colored.
Lamcombe	Canada	---	---	---
Large Black	United Kingdom	---	270-360 kg (600-790 lb)	Black
Large Black-white	---	---	---	Black & White
Large White	United Kingdom	---	---	White
Latvian White	Latvia	---	---	White
Leicoma	---	---	---	---
Lithuanian Native	Lithuania	---	---	Black & White, Red, Black & Tri-Colored
Lithuanian White	Lithuania	---	---	White
Lincolnshire Curly-Coated Pig	United Kingdom	---	---	White
Livny	Russia	---	---	---
Malhado de Alcobaça	Portugal	---	---	Black & White
Mangalitsa	Hungary	---	---	Blonde, Swallow-Bellied, Red
Meishan	China	---	---	Black
Middle White	United Kingdom	---	---	White
Minzhu	---	---	---	---
Minokawa Buta	---	---	---	---
Mong Cai	---	---	---	---
Mora Romagnola	---	---	---	---
Moura	---	---	---	---
Mukota	Zimbabwe	---	---	---
Mulefoot	Gulf Coast	---	---	---
Murom	---	---	---	---
Myrhorod	Ukraine	---	---	---
Neijiang	---	---	---	---

Contd..

Breed	Origin	Height	Weight	Color
Ningxiang	---	---	---	---
North Caucasian	Russia & Uzbekistan	---	---	---
North Siberian	Russia	---	---	---
Norwegian Landrace	Norway	---	---	White
Norwegian Yorkshire	Norway	---	---	---
Ossabaw Island	Ossabaw Island	51 cm (20 in)	90 kg (200 lb)	Black, Spotted
Oxford Sandy and Black	United Kingdom	---	---	Red & Black spotted
Philippine Native	---	---	---	---
Piétrain	Wallonia	---	---	White spotted
Poland China	USA	---	---	---
Red Wattle	USA	120 cm (47 in)	270-680 kg (600-1,500 lb)	Red
Semirechye	Kazakhstan	---	---	---
Siberian Black Pied	Russia	---	---	Black & White
Small Black	United Kingdom	---	---	Black
Small White	United Kingdom	---	---	White
Spots	---	---	---	---
Surabaya Babi	---	---	---	---
Swabian-Hall	Germany	---	---	Black & White
Swedish Landrace	Sweden	---	---	White
Taihu pig	China	---	---	Black
Tamworth	United Kingdom	50-65 cm (20-26 in)	250-370 kg (550-820 lb)	Red
Thuoc Nhieu	---	---	---	---
Tibetan	---	---	---	---
Tokyo-X	Japan	---	---	---
Tsivilsk	Russia	---	---	White
Turopolje	Croatia	---	---	---
Ukrainian Spotted Steppe	Ukraine	---	---	---
Ukrainian White Steppe	Ukraine	---	---	White
Urzhum	Russia	---	---	---
Vietnamese Potbelly	Vietnam	40-66 cm (16-26 in)	45-90 kg (99-200 lb)	Black, Black with white markings
Welsh	United Kingdom	---	110-140 kg (240-310 lb)	White
Wessex Saddleback	United Kingdom	---	---	Black & White
West French White	France	---	---	White
Windsnyer	---	---	---	---
Wuzishan	---	---	---	---
Yanan	---	---	---	---
Yorkshire Blue and White	United Kingdom	---	---	---
Wild boar	---	---	---	---

WORLD PIG POPULATION

World Pig Population by Country - 2002

Country	No. of Head	Country	No. of Head
Armenia	98,000	Azerbaijan	17,670
Albania	110,000	Benin	460,000
Algeria	5,700	Denmark	12,990,000
Amer Samoa	10,700	Dominica	5,000
Angola	800,000	Dominican Rp	565,529
Antigua Barb	5,250	Belarus	3,372,600
Argentina	4,250,000	Ecuador	2,400,000
Australia	2,763,000	Egypt	29,500
Austria	3,440,405	El Salvador	150,000
Bahamas	4,873	Eq Guinea	6,100
Barbados	35,000	Estonia	345,000
Bel-Lux	6,851,000	Falkland Is	30
Bermuda	600	Fiji Islands	137,000
Bhutan	41,400	Finland	1,300,000
Bolivia	2,850,547	France	14,800,000
Botswana	7,000	Fr Guiana	10,500
Brazil	30,000,000	Fr Polynesia	37,000
Belize	24,718	Georgia	465,000
Solomon Is	67,000	Gabon	213,000
Brunei Darsm	6,000	Gambia	7,962
Bulgaria	1,144,000	Germany	25,957,760
Myanmar	4,498,680	Bosnia Herzg	300,000
Burundi	70,000	Ghana	324,000
Cameroon	1,350,000	Kiribati	12,000
Canada	14,367,100	Greece	938,000
Cape Verde	200,000	Grenada	5,800
Cayman Is	399	Guadeloupe	19,000
Cent Afr Rep	680,000	Guam	5,000
Sri Lanka	68,300	Guatemala	1,450,000
Chad	24,000	Guinea	97,835
Chile	2,750,000	Guyana	20,000
China, Main	457,430,000	Haiti	1,000,500
Colombia	2,350,000	Honduras	480,000
Congo, Rep	46,000	China,H.Kong	100,000
Cook Is	40,000	Hungary	4,822,000
Costa Rica	475,000	Croatia	1,300,000
Cuba	2,700,000	Iceland	44,000
Cyprus	445,268	India	17,500,000

Country	No. of Head	Country	No. of Head
Indonesia	6,000,000	Vanuatu	62,000
Ireland	1,762,900	New Zealand	358,068
Israel	155,000	Nicaragua	405,000
Italy	8,410,000	Niger	39,000
Côte dIvoire	336,000	Nigeria	5,100,000
Kazakhstan	1,100,000	Niue	1,700
Jamaica	180,000	Norway	400,000
Japan	9,612,000	Panama	280,000
Kyrgyzstan	80,829	Czech Rep	3,440,925
Kenya	315,000	Papua N Guin	1,650,000
Cambodia	2,114,524	Paraguay	2,750,000
Korea D P Rp	3,137,000	Peru	2,800,000
Korea Rep	8,811,000	Philippines	11,652,700
Latvia	428,700	Poland	18,707,000
Laos	1,425,900	Portugal	2,389,000
Lebanon	64,000	GuineaBissau	350,000
Lesotho	60,000	Timor-Leste	300,000
Liberia	130,000	Puerto Rico	118,000
Liechtensten	3,000	Zimbabwe	604,296
Lithuania	1,011,000	Réunion	77,000
Madagascar	850,000	Romania	3,950,000
Malawi	456,291	Rwanda	180,000
Malaysia	2,100,000	Russian Fed	15,923,000
Mali	85,000	Yugoslavia	3,328,793
Malta	69,000	St Helena	622
Martinique	35,000	St Kitts Nev	4,900
Mauritius	13,500	St Lucia	14,950
Mexico	18,000,000	St Vincent	9,500
Mongolia	15,000	Sao Tome Prn	2,200
Montserrat	1,200	Senegal	280,000
Morocco	8,000	Seychelles	18,500
Mozambique	180,000	Sierra Leone	52,000
Micronesia	32,000	Slovenia	600,000
Moldova Rep	448,900	Slovakia	1,517,000
Namibia	21,854	Singapore	190,000
Nauru	2,800	Somalia	4,000
Nepal	934,461	South Africa	1,600,000
Netherlands	13,000,000	Spain	23,857,780
NethAntilles	2,400	Suriname	24,000
NewCaledonia	40,000	Tajikistan	500
Macedonia	200,000	Swaziland	34,000

Country	No. of Head	Country	No. of Head
Sweden	1,881,743	UK	5,527,000
Switzerland	1,536,000	Ukraine	8,478,000
Syria	770	USA	59,138,000
Turkmenistan	45,000	Burkina Faso	630,000
China, Taiwan	7,165,000	Uruguay	385,000
Tanzania	450,000	Uzbekistan	90,000
Thailand	6,688,904	Venezuela	5,654,968
Togo	289,200	Viet Nam	21,740,700
Tokelau	1,000	Ethiopia	25,000
Tonga	80,853	Br Virgin Is	1,500
Trinidad Tob	42,000	US Virgin Is	2,600
Tunisia	6,000	Wallis Fut I	25,000
Turkey	2,700	Samoa	201,000
Uganda	1,550,000	Congo, Dem R	953,066
Tuvalu	13,200	Zambia	340,000
		World Total	**939,350,700**

DIFFERENT BREEDS RABBITS

Rabbit breeds are different varieties of the *domestic rabbit* created through *selective breeding* or *natural selection*. Breeds recognized by organizations such as the *American Rabbit Breeders' Association* (ARBA) or the *British Rabbit Council* (BRC) may be exhibited and judged in rabbit shows. *Breeders* attempt to emulate the *breed standard* by which each breed is judged.

Breed	Origin	Size	Fur Type	Ear Type	Colour	ARBA Accepted	BRC Accepted
Alaska	Germany	3.2-4.1 kg	Short	Upright	Black	No	Yes
Altex	United States	5.9 kg	Short	Upright	Albino	No	No
American Blue	United States	4.1-5.4 kg	Short	Upright	Blue	Yes	No
American White	United States	4.1-5.4 kg	Short	Upright	White	Yes	No
American Fuzzy Lop	United States	1.4-1.8 kg	Long	Lop	All	Yes	No
American Sable	United States	4.1-4.5 kg	Short	Upright	Sable	Yes	No
Argente Bleu	France	2.7 kg	Short	Upright	Blue	No	Yes
Argente Brun	France	2.7 kg	Short	Upright	Brown	No	Yes
Argente Clair	France	2.7 kg	Short	Upright	Blue Silver	No	No
Argente Crème	France	2.7 kg	Short	Upright	Cream	Yes	Yes
Argente de Champagne	France	3.6-4.5 kg	Short	Upright	Silver	Yes	Yes
Argente Noir	France	2.7 kg	Short	Upright	Dark Silver	No	Yes
Argente St Hubert	France	2.7 kg	Short	Upright	Silvered Black Agouti	No	Yes

Breed	*Origin*	*Size*	*Fur Type*	*Ear Type*	*Colour*	*ARBA Accepted*	*BRC Accepted*
Baladi	Egypt	2.7 kg	Short	Upright	Black, Red, White	No	No
Bauscat	Egypt	3.6 kg	Short	Upright	Albino	No	No
Beige	England/ Netherlands	62.9 kg	Short	Upright	Dark Chamois or Light Sand	No	Yes
Belgian Hare	Belgium	3.6-4.5 kg	Short	Upright	Black, Black, & Tan, Red	Yes	Yes
Beveren	Belgium	2.3-3.6 kg	Short	Upright	Black, Blue, White	Yes	Yes
Blanc de Bouscat	France	5.0-6.8 kg	Short	Upright	White	No	Yes
Blanc de Hotot	France	3.6-5.0 kg	Short	Upright	White, dark rings around eyes	Yes	Yes
Blanc de Popielno	France	4.5-5.4 kg	Short	Upright	White	No	No
Blanc de Termonde	Belgium	4.5-5.4 kg	Short	Upright	White, Red Eyes	No	Yes
Blue of Ham	Belgium	4.5-5.9 kg	Short	Upright		No	No
Blue of Sint -Niklaas	Belgium	2.3-5.4 kg	Short	Upright	Blue	No	No
Bourbonnais Grey	France	.2-5.0 kg	Short	Upright	Blue	No	No
Brazilian	Brazil	3.2-5.0 kg	Short	Upright		No	No
Britannia Petite	England	0.68-1.1 kg	Short	Upright	Black, Black Otter, Chestnut, Sable Marten, Red-Eyed White. Blue-Eyed White	Yes	No
British Giant	England	5.4-5.9 kg	Short	Upright	Black, Blue, White, Opal, Steel Grey, Brown Grey	No	Yes
Brown Chestnut of Lorraine	France	42.0 kg	Short	Upright	Chestnut	No	No
Caldes	Spain		Short	Upright	Red-Eyed White	No	No
Californian	United States	4.1-4.5 kg	Short	Upright	White with Chocolate, Lilac or Blue points, red eyes	Yes	Yes
Carmagnola Grey	Italy	4.5-5.4 kg	Short	Upright	Chinchilla	No	No
Cashmere Lop	England	1.8-2.3 kg	Long	Lop	Many	No	Yes
Chaudry	France	3.6-4.1 kg	Short	Upright	Red-Eyed White	No	No

Contd...

Breed	Origin	Size	Fur Type	Ear Type	Colour	ARBA Accepted	BRC Accepted
Checkered Giant	Europe	5.0-5.4 kg	Short	Upright	White with black markings	Yes	Yes
Chinchilla (Standard)	France	2.3-3.2 kg	Short	Upright	Chinchilla	Yes	Yes
Chinchilla (American)	United States	4.1-5.4 kg	Short	Upright	Chinchilla	Yes	No
Chinchilla (Giganta)	France	4.5-5.4 kg	Short	Upright	Chinchilla	No	Yes
Chinchilla (Giant)	United States	4.5-7.3 kg	Short	Upright	Chinchilla	Yes	No
Cinnamon	United States	4.5-5.0 kg	Short	Upright	Cinnamon	Yes	No
Continental Giant	Europe	5.4-7.3 kg	Short	Upright	Many	No	Yes
Criollo	Mexico		Short	Upright		No	No
Cuban Brown	Central America		Short	Upright	Chocolate	No	No
Czech Albin	Czech Republic		Short	Upright	White	No	No
Czech Spot	Czech Republic	2.7-3.6 kg	Short	Upright	Black, Blue, Agouti, Tri-color	No	No
Czech Red rabbit	Czech Republic	1.8-2.3 kg	Short	Upright	Chestnut	No	No
Deilenaar	Netherlands	2.3-3.6 kg	Short	Upright	Red Agouti	No	Yes
Dutch	England	1.8-2.3 kg	Short	Upright	Black, Blue, Chinchilla, Chocolate, Grey, Steel, Tortoise with a white band across the shoulders, white nose and white paws	Yes	Yes
Dutch (Tri-Coloured)	England	1.8-2.3 kg	Short	Upright	Tri-colour with a white band across the shoulders, white nose and white paws	No	Yes
Dwarf Hotot	Germany	20.91-1.4 kg	Short	Upright	White with black around the eyes	Yes	No
Dwarf Lop (Mini Lop in USA)	Germany	1.8-2.3 kg	Short	Lop	Many	Yes	Yes

Contd...

Breed	Origin	Size	Fur Type	Ear Type	Colour	ARBA Accepted	BRC Accepted
Elfin	Sweden	1.8-2.3 kg	Short	Upright	Many	No	No
Enderby Island	Australia	1.4-1.8 kg	Short	Upright	Champagne, Crème	No	No
English Angora	France	2.3-3.6 kg	Long	Upright	Many	Yes	Yes
English Spot	England	2.3-3.6 kg	Short	Upright	White with colored butterfly pattern	Yes	Yes
English Lop	England	4.5-5.0 kg	Short	Lop	Many	Yes	Yes
Fauve de Bourgogne	France	3.2-5.0 kg	Short	Upright	Orange/Red	No	Yes
Fee de Marbourg (Marburger)	France	1.8-3.2 kg	Short	Upright	Lilac	No	No
Flemish Giant	Belgium	6.4 kg	Short	Upright	Steel, Grey, Sandy	Yes	Yes
Florida White	United States	1.8-2.7 kg	Short	Upright	White	Yes	No
French Angora	France	3.2-4.5 kg	Long	Upright	Many	Yes	No
French Lop	France	4.5 kg	Short	Lop	Many	Yes	Yes
Gabali	Egypt	2.7-3.2 kg	Short	Upright	Agouti	No	No
German Angora	Germany	1.8-5.4 kg	Long	Upright	White, Albino	No	No
German Lop	Germany	2.7-3.6 kg	Short	Lop	Many	No	Yes
Giant Angora	United States	4.5 kg	Long	Upright		Yes	No
Giant Papillon	Europe	5.0-5.4 kg	Short	Upright	White with black markings	Yes	Yes
Giza White	Egypt	2.3-3.2 kg	Short	Upright	White (albino)	No	No
Golden Glavcot	England	2.3-2.7 kg	Short	Upright		No	Yes
Gotland	Sweden	3.0-4.1 kg	Short	Upright		No	No
Grey Pearl of Halle	Belgium	2.3-2.7 kg	Short	Upright	Light Grey	No	No
Güzelçaml? rabbit	Turkey	2.3-3.6 kg	Short	Upright	White with chocolate markings	No	No
Harlequin	France	2.7-4.1 kg		Upright	2 colours in fur i.e. part black part orange	Yes	Yes
Havana	Netherlands	2.0-2.9 kg	Short	Upright	Chocolate, Blue, Black, Lilac	Yes	Yes
Himalayan	Asia	2.7-3.6 kg	Short	Upright	White with black, chocolate, lilac or blue points	Yes	Yes
Hulstlander	Netherlands	1.8-2.7 kg	Short	Upright	White with blue eyes	No	Yes

Contd...

Breed	Origin	Size	Fur Type	Ear Type	Colour	ARBA Accepted	BRC Accepted
Hungarian Giant	Hungary	5.4-6.8 kg	Short	Upright	Many	No	No
Jersey Wooly	United States	1.1-1.6 kg	Long	Upright	Many	Yes	No
Kabyle	Algeria	2.3-2.7 kg	Short	Upright	Many	No	No
Lilac	England	2.3-3.2 kg	Short	Upright	Lavender	Yes	Yes
Lionhead	United States	1.1-1.6 kg	Long	Upright		No	Yes
Liptov Balds-potted Rabbit	Slovakia	3.6-4.1 kg	Short	Upright	Agouti, Black	No	No
Meissner Lop	Germany	4.5-5.4 kg	Short	Lop	Many	No	Yes
Mellerud rabbit	Sweden	3.0-3.5 kg	Short	Upright	Albino, Black/white	No	No
Miniature Lop (Holland Lop in USA)	Netherlands	0.91-1.4 kg	Short	Lop	Many	Yes	Yes
Mini Lion Lop	England	0.91-1.4 kg	Long	Lop	Many	No	Yes
Netherland Dwarf	Netherlands	0.50-1.6 kg	Short	Upright	Many	Yes	Yes
New Zealand	United States	4.1-5.4 kg	Short	Upright	White with red eyes, Red, Black, Broken	Yes	Yes
New Zealand Red	Czech Republic	4.1-5.4 kg	Short	Upright	Chestnut	No	Yes
Orestad	Scandinavia	2.3-3.2 kg	Short	Upright	Albino	No	No
Palomino	United States	4.5-5.0 kg	Short	Upright	Lynx, Golden	Yes	No
Pani	Japan		Short	Upright		No	No
Pannon White	Hungary	2.3-4.5 kg	Short	Upright	White	No	No
Perlfee	England	2.3-3.6 kg	Short	Upright	Blue Agouti	No	Yes
Plush Lop (Standard)	United States	2.3-3.6 kg	Rex	Lop		No	No
Plush Lop (Mini)	United States	1.4-1.8 kg	Rex	Lop		No	No
Pointed Beveren	Belgium	3.2 kg	Short	Upright	Blue points, brown or lilac also recognised	No	Yes
Polish	England	0.45-0.91 kg	Short	Upright	Many	Yes	Yes
Rex (Standard)	France	2.7-4.5 kg	Rex[1]	Upright	Many	Yes	Yes
Rex (Astrex)	England	2.7-4.5 kg	Rex, wavy	Upright	Many	No	Yes
Rex (Mini)	England	1.4-1.8 kg	Rex	Upright	Many	Yes	Yes
Rex (Opossum)	England	2.7-3.6 kg	Long	Upright	Many	No	Yes
Rhinelander	Germany	4.1-4.5 kg	Short	Upright	White with coloured utterfly patterns	Yes	Yes

Contd...

Breed	Origin	Size	Fur Type	Ear Type	Colour	ARBA Accepted	BRC Accepted
Sallander	Netherlands	2.3-4.1 kg	Short	Upright		No	Yes
San Juan	United States	1.4-2.3 kg	Short	Upright	Agouti	No	No
Satin		2.7-3.6 kg	Short	Upright	All self colours, very shiny fur	Yes	Yes
Satin (Mini)		1.8-2.3 kg	Short	Upright		No	Yes
Satin Angora	Canada	2.7-4.1 kg	Long	Upright		Yes	No
Sachsengold	Germany		Short	Upright	Chestnut	No	No
Siberian		2.3-3.2 kg	Short	Upright		No	Yes
Siamese Sable		2.3-3.2 kg	Short	Upright		No	Yes
Silver		2.3-3.2 kg	Short	Upright		Yes	Yes
Silver Fox	United States	5.0-5.4 kg	Short	Upright		Yes	Yes
Silver Marten		4.3 kg	Short	Upright		Yes	No
Smoke Pearl		2.3-3.2 kg	Short	Upright		No	Yes
Spanish Giant	Spain	5.7-6.8 kg	Short	Upright		No	No
Squirrel		2.3-3.2 kg	Short	Upright		No	Yes
Sussex	England	3.2-3.6 kg	Short	Upright	Gold, Cream	No	Yes
Swiss Fox		2.3-3.6 kg	Long	Upright		No	Yes
Tadla	Morocco	1.8-2.3 kg	Short	Upright	Agouti	No	No
Tan	England	1.8-2.7 kg	Short	Upright	Black, Blue, Chocolate, Lilac	Yes	Yes
Teddywidder	Germany	1.5 kg	Long	Lop	Many	No	No
Thrianta	Swiss Alps	2.7 kg	Short	Upright	Chestnut	Yes	Yes
Thuringer	Germany	4.1 kg	Short	Upright		No	Yes
Vienna		3.2-4.1 kg	Short	Upright	Blue, Black, Agouti, White	No	Yes
Wheaten		2.5-3.2 kg	Short	Upright	Fawn or pale yellow	No	Yes
Wheaten Lynx		2.5-3.2 kg	Short	Upright	Orange shot silver	No	Yes
Zemmouri	Morocco		Short	Upright			

THE DIFFERENT REASONS TO BREED RABBITS

People from all around the world breed rabbits for a multitude of different reasons. Here is a list of some of the most popular reasons why people decide to get started breeding rabbits.

- **Rabbits Make Great Pets**: One of the most popular reasons people start breeding rabbits is to raise them up for the large pet market. Since *rabbits* are so cute and flat out adorable they make great pets for children and adults alike. In fact in the UK, the rabbit is the third most popular pet option.

- **Rabbits are a Quiet Species**: Rabbits are in fact a very quiet species which very rarely make a lot of noise. This is a great benefit for urban farmers who want to keep livestock but, don't want to disturb the neighbors and create a scene.

- **Breeding Rabbits is Relatively Low Maintenance**: Once you get in the habit of raising and feeding your rabbits, it's pretty easy to keep the project going, in comparison to the majority of other livestock projects.

- **Rabbit Meat is Healthy**: Thousands, if not millions of people worldwide raise rabbits for meat. Rabbit meat has been proven to be one of the healthiest meats on the planet with its high level of protein and low cholesterol + fat levels.

- **Rabbit Showing is Popular**: Thousands of people in the United States alone focus on raising and breeding rabbits for show. Each year thousands of rabbit shows are held in the continental United States, where rabbits are judged according to ARBA rules and standards. Showing has proven to be a great and enjoyable hobby for many rabbit raisers all over the globe.

- **Breeding Rabbits is Economical**: Breeding rabbits is a very economical project in comparison to many of the other livestock projects that people choose. As long as you know how to effectively raise and sell your rabbits you can keep the costs pretty low. If you do everything right, you might even be able to start generating a profit from your rabbit project.

- **Rabbits are Awesome + Fun to Raise**: Lastly rabbits are an amazing and diverse species that can be fun to raise.

TOP TEN MOST DANGEROUS DOG BREEDS

1. **Pit-bull:** The Pit-bull Dog breed comprises of different dog breeds namely-American Pit Bull Terrier, the American Staffordshire Terrier, the Staffordshire Terrier and any crosses between the three dog breeds. The Pit-bull Dogs are also used in lots of illegal Dog Fighting in USA. Pit Bull breeds have become famous for their roles as soldiers in the army by helping the Human soldiers, as police dogs, and as search and rescue dogs, actors, television personalities, Seeing Eye dogs and celebrity pets. A pit bull is a fearless dog that will take on any opponent. They will lock their jaws onto the prey until it's dead.

2. **Rottweilers:** Rottweiler Dogs are from medium to large size breed of domestic dog that originated in Rottweil, Germany, The dogs were known as "Rottweil butchers' dogs" because they were used to herd livestock and pull carts laden with butchered meat and other products to market. Rottweilers Dogs are now used as search and rescue dogs, as guide dogs

for the blind, as guard dogs or police dogs. They are an extremely intelligent breed That's why they make great guard dogs, but poor training can lead them to become very aggressive and disobedient. Pit bull and Rottweiler are responsible for an estimated 60 percent of all dog bite fatalities.

3. **Siberian Husky:** Siberian Husky is a medium-size, dense-coat working dog breed that originated in north-eastern Siberia. The breed belongs to the Spitz genetic family, Huskies are active, energetic, and resilient breed, Siberian Huskies were bred by the Chukchi of Northeastern Asia to pull heavy loads long distances through difficult conditions, Because they were originally bred for work and typically did not socialize with humans, this large breed can have behavioral issues and sometimes mistaken children for prey.

4. **Alaskan Malamutes:** Alaskan malamute is a generally large breed of domestic dog. They are sometimes mistaken for a Siberian Husky due to color and markings, but they are quite different in many ways including size, structure and personality. As pets, once mature, Alaskan Malamutes have a very quiet, dignified temperament and are loyal to their owners.

5. **German Shepherd:** The German Shepherd Dog, also known as Alsatian or just the German Shepherd, is a breed of large-sized dog . German Shepherd is a relatively new breed of dog, with its origin dating to 1899. As part of the Herding Group, the German Shepherd is a working dog developed originally for herding and guarding sheep. Because of its strength, intelligence and abilities in obedience training it is widely used by police and military, German shepherds are very intelligent and friendly. But their jaw strength and natural instincts make them equally dangerous in certain situations.

6. **Great Dane:** The Great Dane, also known as German Mastiff or Danish Hound, is a breed of domestic dog known for its giant size. The Great Dane is one of the world's tallest dog breeds, but they are also known for developing aggressive behavior with poor training. At 150 lbs with inches-long teeth, an aggressive dog presents a very dangerous scenario, especially around children.

7. **Doberman Pinscher:** Doberman Pinscher or simply called as Doberman, is a breed of domestic dog originally developed around 1890 by Karl Friedrich Louis Dobermans. Doberman Pinschers are one of the most common Dog Breed, and the breed is well-known as an intelligent, alert, and loyal companion dog. They are known for responding aggressively to physical corrections in training, therefore they can be dangerous.

8. **Chow Chow:** Chows are extremely territorial. Any strangers entering the property or approaching family members are considered a threat by this breed, and even well-trained dogs are known to aggressively defend their territory, Sometimes can be too much dangerous.

9. **Presa Canario:** The Perro de Presa Canario is a large Molosser-type dog breed originally bred for working livestock. The name of the breed is Spanish, means "Canarian blood hound," and is often shortened to "Presa Canario" or simply "Presa." Originally bred to guard and fight with cattle, an attack by this dog has been described as hopeless for the victim. They are a guardian breed with man-stopping ability, incredible power and a complete lack of fear.

10. **Boxer:** Boxer is a breed of stocky, medium-sized, short-haired dog. The coat is smooth and tight-fitting; colors are fawn or brindled, with or without white markings, which may cover the entire body. Boxers are the seventh most popular breed of dog in the United States. These dogs are not typically aggressive by nature. They are bright, energetic and playful breed. Boxers have been known to be "headstrong", which makes it a bit difficult to train them but with positive reinforcement techniques, Boxers often respond much better.

DIFFERENT INDIAN RUNNER DUCKS

Indian Runners (*Anas platyrhynchos domesticus*) are an unusual breed of domestic duck. They stand erect like penguins and, rather than waddling, they run. The females usually lay about 150 – 200 eggs a year or more; depending whether they are from exhibition or utility strains. They were found on the Indonesian Islands of Lombok, Java and Bali where they were 'walked' to market and sold as egg-layers or for meat. These ducks do not fly and only rarely form nests and incubate their own eggs. They run or walk, often dropping their eggs wherever they happen to be. Duck-breeders need to house their birds overnight or be vigilant in picking up the eggs to prevent them from being taken by other animals. The ducks vary in weight between 1.4 and 2.3 kg (3-5 lbs). Their height (from crown to tail tip) ranges from 50 cm (20 inches) in small females to about 76 cm (30 inches) in the taller males. The eggs are often greenish-white in color, but these too vary. Indian Runners love foraging. They also like swimming in ponds and streams, but they are likely to be preoccupied in running around grassy meadows looking for worms, slugs, even catching flies. They appreciate open spaces but are happy in gardens from which they cannot fly and where they make much less noise than call ducks. Only the females quack. All drakes are limited to a hoarse whisper. Runners eat less in the way of grain and pellet supplement than big table ducks.

This is list of officially recognized duck breeds

1. Abacot Ranger (also known as Streicher)	Call Duck	French White Duck	Magpie Duck	Tsaiya Duck
2. Alabio Duck	Challans	German Pekin Duck	Majorcan Duck	Ukrainian Duck
3. Allier Duck	Campbell Duck	Gimbsheimer Duck	Moc Duck	Venetian Duck
4. American Pekin Duck	Cayuga Duck	Golden Cascade	Mulard	Vit Bau Ben Duck
5. Ancona duck	Co Duck	Gressingham Duck (Wild Mallard crossed with Pekin)	Muscovy Duck	Vouille Duck
6. Australian Spotted	Crested Duck	Havanna Duck	Overberg Duck	Welsh Harlequin Duck
7. Aylesbury Duck	Danish Duck	High Flyer Duck	Pomeranian Duck	
8. Bac Kinh Duck	Dendermond Duck	Hoa Duck	Rouen Duck	
9. Bali duck	Duclair Duck	Hungarian Duck	Saxony Duck	
10.Bashkir Duck	Dutch Duck	Huttengem Duck	Semois	
11.Bau Quy Duck	Dutch Hookbill Duck	Challans Duck	Shetland duck	
12.Black East Indian Duck	East Indie Duck	Chara Chamble	Silver Appleyard Duck	
13.Blekinge duck	Estaires Duck	Chara Chemballi Duck	Swedish Blue duck	
14.Bourbon Duck	Faroese Duck	Indian Runner Duck	Swedish Yelow Duck	
15.Buff Orpington Duck	Forest Duck	Khaki Compbell	Termonde Duck	

DIFFERENT TYPES OF FISH FOUND IN INDIA

1. ***Trichogaster lalius:*** Found in surface area of all finding fresh water resources.

2. ***Danio rerio:*** Usually found during rainy season in the paddy fields and ditches.

3. ***Trichogaster fasciatus:*** Found in surface area of all finding fresh water resources.

4. ***Mastacembelus armatus:*** Found in muddy places.

A. Indigenous major carps

1. ***Catla catla:*** Found on the surface of lake, ponds, tanks, dams and all others fresh water resources.

2. ***Lebeo rohita:*** Found on the column of lakes, ponds, tanks, dams and all others fresh water resources.

3. ***Cirrhinus mrigala:*** Found at the bottom of lakes, ponds tanks, dams etc.

4. ***Lebeo bata:*** Found at the bottom of different water resources.

5. ***Labeo calbasu:*** Found at the bottom region of different fresh water resources.

B. Exotic carps

1. ***Hypophthalmichthys molitrix:*** Found on the surface of water resources.

2. ***Cyprinus carpio var. specularis:*** Found at the bottom region of water resources.

3. ***Cyprinus carpio var. communis:*** Found at the bottom region of different water resources.

4. ***Ctenopharyngodon idella:*** Found in column region of different fresh water resources.

5. ***Tilapia mossambicus:*** Dwelling at the bottom region of different water resources.

C. Air breathing fishes

1. ***Clarias batrachus:*** Found in muddy places and at the bottom of fresh water resources.

2. **Heteropneustes fossilis:** Found in muddy places and at the bottom of fresh water resources.

3. ***Channa punctatus:*** Specially found in muddy places during rainy season and in the bottom of fresh water resources.

4. ***Channa gachua:*** Found in muddy places and in the bottom region of different water resources.

5. ***Channa striatus:*** Found in muddy places, at the bottom of different water resources..

6. ***Channa marulius*** Found in muddy places in least number, at the bottom region of Ponds, Lakes and Dam.

7. ***Wallago attu:*** Found in column region of Ponds, Lake, and Dam.

8. ***Ompok bimaculatus:*** Found in column region of different fresh water resources.

9. ***Notopterus notopterus***

10. ***Notopterus chitala***

11. *Anabas testudineus:* Found in column region of different fresh water resources.

12. *Mystus seenghala:* Found in bottom region of different fresh water resources.

D. Hill stream fishes

1. *Garra gotyla:* Found in river Chano near Churchu block and Konar river.

2. *Garra kempi:* Hill stream of river. Found in ditches near hill foots.

3. *Lepidocephalichthys guntea:* Most commonly occurring hill stream fish.

4. *Nemacheilus botia:* Found in ditches near the foot hill.

E. Common type minor carps

1. *Oxygaster bacaila:* Found in surface area of different fresh water resources.

2. *Amblypharyngodon mola:* Found in bottom region of all finding resources.

3. *Esomus danricus*

4. *Cirrhina reba:* Found in bottom region of different fresh water resources.

5. *Rhinomugil corsula*

6. *Eutropiichthys vacha*

7. *Puntius ticto:* Found in surface area of different water resources.

8. *Puntius stigma:* column region of different fresh water resources.

9. *Puntius conchonius:* Found in column region of all finding fresh water resources.

10. *Ambassis nama:* Found in surface area of Ponds, Lakes, River and Dam.

11. *Ambassis ranga:* Found in surface area of finding water resources.

12. *Glossogobius giuris:* Found in surface area of different finding water resources.

13. *Osteobramo cotio:* Found in surface area of finding fresh water resources.

14. *Barilius bola:* Found in surface area of pond.

15. *Mastacembelus aculeatus:* Found in muddy places.

16. *Mastacembelus pancalus:* Found in muddy places

17. *Mystus bleekeri:* Found in muddy places.

18. *Mystus cavasius:* Found in muddy places.

19. *Mystus vittatus:* Found in muddy places.

F. Rare type minor carps

1. *Guludia chapra:* Found in surface area of pond.
2. *Xenentodon cancila*
3. *Ailia coila*

FISHING BUSINESS IN INDIA

Fishing in India is a major industry in its coastal states, employing over 14 million people. Fish production in India has increased more than tenfold since its independence in 1947. According to the Food and Agriculture Organization (FAO) of the United Nations, fish output in India doubled between 1990 and 2010. India has 8,118 kilometers of marine coastline, 3,827 fishing villages, and 1,914 traditional fish landing centers. India's fresh water resources consist of 195,210 kilometers of rivers and canals, 2.9 million hectares of minor and major reservoirs, 2.4 million hectares of ponds and lakes, and about 0.8 million hectares of flood plain wetlands and water bodies. As of 2010, the marine and freshwater resources offered a combined sustainable catch fishing potential of over 4 million metric tonnes of fish. In addition, India's water and natural resources offer a tenfold growth potential in aquaculture (farm fishing) from 2010 harvest levels of 3.9 million metric tonnes of fish, if India were to adopt fishing knowledge, regulatory reforms, and sustainability policies adopted by China over the last two decades. The marine fish harvested in India consist of about 65 commercially important species/groups. Pelagic and mid water species contributed about 52% of the total marine fish in 2004.

India is a major supplier of fish in the world. In 2006 the country exported over 600,000 metric tonnes of fish, to some 90 countries, earning over $1.8 billion. Shrimps are one of the major varieties exported. The giant tiger prawn (Penaeus monodon) is the dominant species chosen for aquaculture, followed by the Indian white prawn (Fenneropenaeus indicus). Shrimp production from coastal aquaculture during 2004 stood at approximately 120,000 tonnes. Farmed shrimp accounted for about 60% of shrimp exported from the country.

Marine and freshwater catch fishing combined with aquaculture fish farming is a rapidly growing industry in India. In 2008 India was the sixth largest producer of marine and freshwater capture fisheries, and the second largest aquaculture farmed fish producer in the world. Fish as food—both from fish farms and catch fisheries offers India one of the easiest and fastest way to address malnutrition and food security. Despite rapid growth in total fish production, a fish farmers' average annual production in India is only 2 metric tonnes per person, compared to 172 tonnes in Norway, 72 tonnes in Chile, and 6 tonnes per fisherman in China. Higher productivity, knowledge transfer for sustainable fishing, continued growth in fish production with increase in fish exports have the potential for increasing the living standards of Indian fishermen. As of 2010, fish harvest distribution was difficult within India because of poor rural road infrastructure, lack of cold storage and absence of organized retail in most parts of the country.

History

Fishing and aquaculture in India has a long history. Kautilya's Arthashastra (321–300 B.C.) and King Someswara's Manasottara (1127 A. D.) each refer to fish culture. For centuries, India has had a traditional practice of fish culture in small ponds in Eastern India. Significant advances in productivity were made in the state of West Bengal in the early nineteenth century with the controlled breeding of carp in *Bundhs* (tanks or impoundments where river conditions are simulated). Fish culture received notable attention in Tamil Nadu (formerly the state of Madras) as early as 1911, subsequently, states such as West Bengal, Punjab, Uttar Pradesh, Gujarat, Karnataka and Andhra Pradesh initiated fish culture through the establishment of Fisheries Departments. In 2006, Indian central government initiated a dedicated organization focused on fisheries, under its Ministry of Agriculture.

Brackish water farming in India is also an age old system confined mainly to the *Bheries* (manmade impoundments in coastal wetlands) of West Bengal and *pokkali* (salt resistant deepwater paddy) fields along the Kerala coast. With no additional knowledge and technology input, except that of trapping the naturally bred juvenile fish and shrimp seed, these systems have been sustaining production levels of between 500 and 750 kg/ha/year with shrimp contributing 20 to 25 percent of the total Indian production.

Growth

It rose from only 800,000 tons in FY 1950 to 4.1 million tons in the early 1990s. From 1990 through 2010, Indian fish industry growth has accelerated, reaching a total marine and freshwater fish production to about 8 million metric tons. Special efforts have been made to promote extensive and intensive inland fish farming, modernize coastal fisheries, and encourage deep-sea fishing through joint ventures. These efforts led to a more than fourfold increase in coastal fish production from 520,000 tons in five years 1950 to 2.4 million tons in five years 1990. The increase in inland fish production was even more dramatic, increasing almost eightfold from 218,000 tons in five years 1950 to 1.7 million tons in five years 1990. The value of fish and processed fish exports increased from less than 1 percent of the total value of exports in five years 1960 to 3.6 percent in five years 1993.

Economic benefits

Fishing in India contributed over 1 percent of India's annual gross domestic product in 2008. Catch fishing in India employs about 14.5 million people. The country's rich marine and inland water resources, fisheries and aquaculture offer an attractive and promising sector for employment, livelihood, and food security. Fish products from India are well received by almost half of world's countries, creating export-driven employment opportunities in India and greater food security for the world. During the past decades the Indian fisheries and aquaculture has witnessed improvements in craft, tackle and farming methods Creation of required harvest and post-harvest infrastructure has been receiving due attention of the central and state governments. All this has been inducing a steady growth. To harvest the economic benefits from fishing, India is adopting exclusive economic zone, stretching 200 nautical miles (370 km) into the Indian Ocean, encompasses more than 2 million square kilometers. In the mid-1980s, only about 33 percent of that area was being exploited. The potential annual catch from the area has been estimated at 4.5 million tons. In addition to this marine zone, India has about 14,000 km² of brackish water available for aquaculture, of which only 600 km² were being farmed in the early 1990s; about 16,000 km² of freshwater lakes, ponds, and swamps; and nearly 64,000 kilometers of rivers and streams. In 1990, there were 1.7 million full-time fishermen, 1.3 million part-time fishermen, and 2.3 million occasional fishermen, many of whom worked as salt makers, ferrymen, or seamen, or operated boats for hire. In the early 1990s, the fishing fleet consisted of 180,000 traditional craft powered by sails or oars, 26,000 motorized traditional craft, and some 34,000 mechanized boats.

Aquaculture

India laid the foundation for scientific carp farming in the country between 1970 and 1980, by demonstrating high production levels of 8 to 10 tonnes/ hectare/year in an incubation center. The late 1980s saw the dawn of aquaculture in India and transformed fish culture into a more modern enterprise. With economic liberalization of early 1990s, fishing industry got a major investment boost. India's breeding and culture technologies include primarily different species of carp; other species such as catfish, Murrells and prawns are recent additions. The culture systems adopted in the country vary greatly depending on the input available in any particular region as well as on the investment capabilities of the farmer. While extensive aquaculture is carried out in comparatively large water bodies with stocking of the fish seed as the only input beyond utilizing natural productivity, elements of fertilization and feeding have been introduced into semi-intensive culture. The different culture systems in Indian practice include:

- Intensive pond culture with supplementary feeding and aeration (10–15 tonnes/ha/yr)

- Composite carp culture (4–6 tonnes/ha/yr)

- Weed-based carp polyculture (3–4 tonnes/ha/yr)

- Integrated fish farming with poultry, pigs, ducks, horticulture, etc. (3–5 tonnes/ha/yr)

- Pen culture (3–5 tonnes/ha/yr)

- Cage culture (10–15 kg/m²/yr)

- Running-water fish culture (20–50 kg/m²/yr)

Aquaculture resources in India include 2.36 million hectares of ponds and tanks, 1.07 million hectares of *beels, jheels* and derelict waters plus in addition 0.12 million kilometers of canals, 3.15 million hectares of reservoirs and 0.72 million hectares of upland lakes that could be utilized for aquaculture purposes. Ponds and tanks are the prime resources for freshwater aquaculture in India. However, less than 10 percent of India's natural potential is used for aquaculture currently. The FAO of the United Nations estimates that about 1.2 million hectares of potential brackish water area available in India is suitable for farming, in addition to this, around 8.5 million hectares of salt affected areas are also available, of which about 2.6 million hectares could be exclusively utilized for aquaculture due to the unsuitability of these resources for other agriculture based activities. However, just like India's fresh water resources, the total brackish water area under cultivation is only just over 13 percent of the potential water area available. India offers opportunities for highly productive farming of shrimp in its brackish water resources. Carp hatcheries in both the public and private sectors have contributed towards the increase in seed production from 6321 million fry in 1985–1986 to over 18500 million fry in 2007. There are 35 freshwater prawn hatcheries in the coastal states producing over 200 million seed per annum. Furthermore, the 237 shrimp hatcheries with a production capacity of approximately 11.425 billion post larvae per year are meeting the seed requirement of the brackish water shrimp farming sector.

Freshwater aquaculture activity is prominent in the eastern part of the country, particularly the states of West Bengal, Orissa and Andhra Pradesh with new areas coming under culture in the states of Punjab, Haryana, Assam and Tripura. Brackish water aquaculture is mainly concentrated on the coasts of Andhra Pradesh, Tamil Nadu, Orissa and West Bengal. With regards to the market, while the main areas of consumption for freshwater fish are in West Bengal, Bihar, Orissa and northeast India, cultured brackish water shrimps supply India's fish export industry.

Distribution of fish industry in Indian States

Fishing is a diverse industry in India. The table below presents the top ten fish harvesting states in India, for the 2007-2008 agriculture years.

Leading fish producing states in India, 2007–2008

Rank	State	Total production (metric tonnes)
1	West Bengal	1,447,260
2	Andhra Pradesh	1,010,830
3	Gujarat	721,910
4	Kerala	667,330
5	Tamil Nadu	559,360
6	Maharashtra	556,450
7	Orissa	349,480
8	Uttar Pradesh	325,950
9	Bihar	319,100
10	Karnataka	297,690

Between 2000 and 2010, the freshwater prawn farming in India has grown rapidly. The state of Andhra Pradesh dominates the sector with over 86 percent of the total production in India with approximately 60 percent of the total water area dedicated to prawn farming, followed by West Bengal. Mixed farming of freshwater prawn along with carp is also very much accepted as a technologically sound culture practice and a viable option for enhancing farm income. Thirty five freshwater prawn hatcheries, at present producing about 200 million seed per annum, cater for the requirements of the country.

Laws and Regulations

India has a federal structure of government. According to India's constitution, the power of enacting laws is split between India's central government and the Indian states. The state legislatures of India have the power to make laws and regulations with respect to a number of subject-matters, including water (i.e., water supplies, irrigation and canals, drainage and *embankments*, water storage and water power), land (i.e., rights in or over land, land tenure, transfer, and alienation of agricultural land), fisheries, as well as the preservation, protection and improvement of stock and the prevention of animal disease. There are many laws and regulations that may be relevant to fisheries and aquaculture adopted at state level.

At the central level, several key laws and regulations are relevant to fisheries and aquaculture. These include the British-era Indian Fisheries Act (1897), which penalizes the killing of fish by poisoning water and by using explosives; the Environment (Protection) Act (1986), being an umbrella act containing provisions for all environment related issues affecting fisheries and aquaculture industry in India. India also has enacted the Water (Prevention and Control of Pollution) Act (1974) and the Wild Life Protection Act (1972). All these legislations must be read in conjunction with one another, and with the local laws of a specific state, to gain a full picture of the law and regulations that are applicable to fisheries and aquaculture in India.

Research and Trainings

Fisheries research and training institutions are supported by central and state governments that deserve much of the credit for the expansion and improvements in the Indian fishing industry. The principal fisheries research institutions, all of which operate under the *Indian Council of Agricultural Research*, are the *Central Marine Fisheries Research Institute* at *Kochi* (formerly Cochin),*Kerala*; the Central Inland Fisheries Institute at *Barrackpore, West Bengal*; and the *Central Institute of Fisheries Technology* at *Willingdon Island* near Kochi. Most fishery training is provided by the Central Institute for Fishery Education in *Mumbai*, which has ancillary institutions in Barrackpore, *Agra* (*Uttar Pradesh*), and *Hyderabad* (*Andhra Pradesh*). The Central Fisheries Corporation in Calcutta is instrumental in bringing about improvements in fishing methods, ice production, processing, storing, marketing, and constructing and repairing fishing vessels. Operating under a 1972 law, the Marine Products Export Development Authority (MPEDA), headquartered in Kochi, has made several market surveys abroad and has been instrumental in introducing and enforcing hygiene standards that have gained for Indian fishery export products a reputation for cleanliness and quality.

Programmes

The Government of India launched National Fisheries Development Board in 2006. Its headquarters are in Hyderabad, located in a fish shaped building. Its activity focus areas are:

- Intensive Aquaculture in Ponds and Tanks
- Fisheries Development in Reservoirs
- Coastal Aquaculture
- Mariculture
- Seaweed Cultivation
- Infrastructure: Fishing Harbours and Landing Centers
- Fish Dressing Centers and Solar Drying of Fish
- Domestic Marketing
- Technology Upgradation
- Deep Sea Fishing and Tuna Processing

The implementation of two programs for inland fisheries establishing fish farmers' development agencies and the National Programme of Fish Seed Development has led to encouragingly increased production, which reached 1.5 million tons during five years 1990, up from 0.9 million tons in five years 1984. A network of 313 fish farmers' development agencies was functioning in 1992. Under the National Programme of Fish Seed Development, forty fish-seed hatcheries were commissioned. Fish-seed production doubled from 5 billion fry in five years 1983 to 10 billion fry in five years 1989. A new program using organic waste for aquaculture was started in five years 1986. Inland fish production as a percent of total fish production increased from 36 percent in five years 1980 to 40 percent by five years 1990.

Rajiv Gandhi Centre for Aquaculture (RGCA)

Rajiv Gandhi Centre for Aquaculture is the Research and Development arm of the Marine Products Export Development Authority (MPEDA), which inspired by the late Prime Minister Rajiv Gandhi's vision of making India a technologically advanced nation, founded this Centre of Excellence in Aquaculture and dedicated it to the development of the Indian Aquaculture Industry. RGCA is actively involved in the development of various Sustainable Aquaculture Technologies that are bio-secure, eco-friendly, traceable and with low carbon outputs, for seed production and grow out farming of various aquatic species, those having export potential in particular. RGCA is also developing a state-of-the-art technology transfer and training centre for disseminating the technologies developed at the various projects established at different locations in the country to the aquaculture industry in India.

Institutions working on fishing business

There are several specialized institutes that train fishermen. The Central Institute of Fisheries Nautical and Engineering Training in Juhu instruct operators of deep-sea fishing vessels and technicians for shore establishments. It has facilities in Madras and Vishakhapatnam for about 500 trainees a year. An Institute named "Fisheries Institute of Technology and Training" (FITT) was established with the participation of TATAs in Tamil Nadu, to improve the socioeconomic condition of fishers. The Integrated Fisheries Project, also headquartered in Kochi, was established for the processing, popularizing, and marketing of unusual fish. Another training organization, the Central in Bangalore, has done techno-economic feasibility studies on locations of fishing harbor sites and brackish-water fish farms. At present there are 19 Fisheries colleges and one fisheries university (CIFE: Central Institute of Fisheries Education, Mumbai) functioning in various states of the country, providing Professional Fisheries education with a view of developing Professionalism in the field of Fisheries. Among the fisheries colleges, Fisheries college and Research Institute located in Tuticorin, Tamil Nadu is the more popular college because of the maximum number of intake of MFSc and PhD candidates every year. Other colleges such as the College of Fisheries, Panangad, College of Fisheries, and Mangalore are also working well for the professionalism.

DIFFERENT FOODS PRODUCTS PREPARED FROM FISH

Fish and fish products are consumed as food all over the world. With other sea foods, it provides the world's prime source of high-quality protein: 14–16 percent of the animal protein consumed worldwide. Over one billion people rely on fish as their primary source of animal protein. Fish and other aquatic organisms are also processed into various food and non-food products.

Processed fish products

- Surimi refers to a Japanese food product intended to mimic the meat of lobster, crab, and other shellfish. It is typically made from white-fleshed fish (such as pollock or hake) that has been pulverized to a paste and attains a rubbery texture when cooked.

- Fish glue is made by boiling the skin, bones and swim bladders of fish. Fish glue has long been valued for its use in all manner of products from illuminated manuscripts to the Mongolian war bow.

- Fish oil is recommended for a healthy diet because it contains the omega-3 fatty acids, eicosapentaenoic acid (EPA), and docosahexaenoic acid (DHA), precursors to eicosanoids that reduce inflammation throughout the body.

- Fish emulsion is a fertilizer emulsion that is produced from the fluid remains of fish processed for fish oil and fish meal industrially.

- Fish hydrolysate is ground up fish carcasses. After the usable portions are removed for human consumption, the remaining fish body – guts, bones, cartilage, scales, meat, etc. – are put into water and ground up.

- Fish meal is made from both whole fish and the bones and offal from processed fish. It is a brown powder or cake obtained by rendering pressing the whole fish or fish trimmings to remove the fish oil. It used as a high-protein supplement in aquaculture feed.

- Fish sauce is a condiment that is derived from fish that have been allowed to ferment. It is an essential ingredient in many curries and sauces.

- Isinglass is a substance obtained from the swim bladders of fish (especially sturgeon), it is used for the clarification of wine and beer.

- Tatami iwashi is a Japanese processed food product made from baby sardines laid out and dried while entwined in a single layer to form a large mat-like sheet.

Fish oil

Fish oil is oil derived from the tissues of oily fish. Fish oils contain the omega-3 fatty acids eicosapentaenoic acid (EPA) and docosahexaenoic acid

(DHA), precursors of certain eicosanoids that are known to reduce inflammation in the body, and have other health benefits. Fish do not actually produce omega-3 fatty acids, but instead accumulate them by consuming either microalgae or prey fish that have accumulated omega-3 fatty acids, together with a high quantity of antioxidants such as iodide and selenium, from microalgae, where these antioxidants are able to protect the fragile polyunsaturated lipids from peroxidation.

Fatty predatory fish like sharks, swordfish, tilefish, and albacore tuna may be high in omega-3 fatty acids, but due to their position at the top of the food chain, these species can also accumulate toxic substances through biomagnification. For this reason, the U.S. Food and Drug Administration recommends limiting consumption of certain (predatory) fish species (e.g. albacore tuna, shark, king mackerel, tilefish and swordfish) due to high levels of toxic contaminants such as mercury, dioxin, PCBs and chlordane.[6] Fish oil is used as a component in aquaculture feed. More than 50 percent of the world's fish oil used in aquaculture feed is fed to farmed salmon. Marine and freshwater fish oil varies in contents of arachidonic acid, EPA and DHA. The various species range from lean to fatty and their oil content in the tissues has been shown to vary from 0.7–15.5%. They also differ in their effects on organ lipids. Studies have revealed that there is no relation between total fish intake and estimated omega"3 fatty acid intake from all fish and serum omega"3 fatty acid concentrations. Only fatty fish intake, particularly salmonid, and estimated EPA + DHA intake from fatty fish has been observed to be significantly associated with increase in serum EPA + DHA. The omega-3 fatty acids in fish oil are thought to be beneficial in treating hyper triglyceridemia, and possibly beneficial in preventing heart disease. Fish oil and omega-3 fatty acids have been studied in a wide variety of other conditions, such as clinical depression, anxiety, cancer, and macular degeneration, although benefit in these conditions remains to be proven.

Fish meal

Fish meal, or fishmeal, is a commercial product made from fish and the bones and offal from processed fish. It is a brown powder or cake obtained by drying the fish or fish trimmings, often after cooking, and then grinding it. If it is a fatty fish it is also pressed to extract most of the fish oil. Fishmeal is a nutrient-rich and high protein supplement feed ingredient that stores well, and is used primarily in diets for domestic animals and sometimes as a high-quality organic fertilizer.

Raw material

Fishmeal can be made from almost any type of seafood but is generally manufactured from wild-caught, small marine fish that contain a high percentage of bones and oil, and is usually deemed not suitable for direct human consumption. The fish caught for fishmeal purposes solely are termed "industrial". Other sources of fishmeal are from by catch of other fisheries and by-products of trimmings made during processing (fish waste or offal) of various seafood products destined for direct human consumption. Virtually any fish or shellfish in the sea can be used to make fishmeal, although there may be a few rare unexploited species which would produce a poisonous meal.

Selection of species

1. The species must be in large concentrations to give a high catching rate; this is essential because the value of industrial fish is less than that of fish for direct human consumption.

2. The fishery should preferably be based on more than one species in order to reduce the effect of fluctuations in supply of any one species.

3. The total abundance of long-lived species varies less from year to year, and 4. Species with a high fat content are more profitable, because the fat in fish is held at the expense of water and not at the expense of protein

Fish meal is manufactured primarily from anchovies in Peru; menhaden in the United States; pout in Norway; capelin, sand eel and mackerel in other parts of northern Europe; and sauries, mackerels and sardines in Japan.

Production of fish meal

Fishmeal is made by either cooking, pressing, drying and grinding of fish or fish waste to which no other matter has been added. It is a solid product from which most of the water is removed and some or all of the oil is removed. Four or five tonnes of fish are needed to manufacture one tonne of dry fishmeal.

There are several ways of making fishmeal from raw fish; the simplest way is to let the fish dry out in the sun. This method is still used in some parts of the world where processing plants are not available; nevertheless the end product is poor in comparison with ones made by modern methods. Nowadays all industrial fish meal is made by the following processes:

Cooking: A commercial cooker is a long steam jacketed cylinder through which the fish are moved by a screw conveyor. This is a critical stage in preparing the fishmeal, as incomplete cooking means that the liquor from the fish cannot be pressed out satisfactorily and overcooking makes the material too soft for pressing. No drying occurs in the cooking stage.

Pressing: A perforated tube with increasing pressure is used for this process. This stage involves removing some of the oil and water from the material and the solid is known as press cake. The water content in pressing is reduced from 70% to about 50% and oil is reduced to 4%.

Drying: It is important to get this stage of the process right. If the meal is under-dried, moulds or bacteria may grow. If it is over-dried, scorching may occur and this reduces the nutritional value of the meal.

Two main types of dryer: Direct and Indirect

Direct: Very hot air at a temperature of 500 °C (932 °F) is passed over the material as it is tumbled rapidly in a cylindrical drum. This is the quicker method, but heat damage is much more likely if the process is not carefully controlled.

Indirect: Cylinder containing steam heated discs which also tumble the meal.

Grinding: This is the last step in processing which involves the breakdown of any lumps or particles of bone.

Nutritional Composition

Any complete diet must contain some protein, but the nutritional value of the protein relates directly to its amino acid composition and digestibility. The amino acid profile of fishmeal is what makes this feed ingredient so attractive as a protein supplement. High-quality fishmeal normally contains between 60% and 72% crude protein by weight. Typical diets for fish may contain from 32% to 45% total protein by weight. Fishmeal is sought after as an ingredient in aquaculture diets because it contains compounds that make the feed more palatable. This allows the feed to be ingested rapidly, and will reduce nutrient leaching. It is thought the non-essential amino acid glutamic acid is one of the compounds that impart palatability to fishmeal.

Fish lipids are also highly digestible by all species of animals and are excellent sources of the essential polyunsaturated fatty acids (PUFA) in both the omega-3 and omega-6 families of fatty acids. The predominant omega-3 fatty acids in fishmeal and fish oil are linolenic acid, docosahexaenoic acid (DHA), and eicosapentaenoic acid (EPA). Essential fatty acids are necessary for normal larval development, fish growth, and reproduction. They are important in normal development of the skin, nervous system, brain, and visual acuity. PUFAs appear to assist the immune system in defense of disease agents and reduce the stress response. Fishmeal also contains valuable phospholipids, fat-soluble vitamins, and steroid hormones. Such high digestibility of fish lipids means they can provide lots of usable energy. If a diet does not provide enough energy, the fish or shrimp will have to break down valuable protein for energy, which is expensive and can increase production of toxic ammonia. Fishmeal is considered to be a moderately rich source of vitamins of the B-complex especially cobalamine (B12), niacin, choline, pantothenic acid, and riboflavin.

Benefits of Fish Meal

Fishmeal in diets increase feed efficiency and growth through better food palatability and enhances nutrient uptake, digestion and absorption. The balanced amino acid composition of fishmeal complements and provides synergistic effects with other animal and vegetable proteins in the diet to promote fast growth and reduce feeding costs. High quality fishmeal provides a balanced amount of all essential amino aids, phospholipids and fatty acids required for optimum development, growth and reproduction especially of larvae and brood stock. The nutrients in fishmeal also aid in disease resistance by boosting and helping to maintain a healthy functional immune system. It also allows for formulation of nutrient-dense diets, which promote optimal growth. Incorporation of fishmeal into diets of aquatic animals helps to reduce pollution from the waste water effluent by providing greater nutrient digestibility. The incorporation of high-quality fishmeal into feed imparts a 'natural or wholesome' characteristic to the final product, such as that provided by wild fish.

World Fish Production

According to the Food and Agriculture Organization (FAO), the world harvest in 2005 consisted of 93.2 million tonnes captured by commercial fishing in wild fisheries, plus 48.1 million tonnes produced by fish farms. In addition, 1.3 million tons of aquatic plants (seaweed etc.) were captured in wild fisheries and 14.8 million tons were produced by aquaculture. The number of individual fish caught in the wild has been estimated at 0.97-2.7 trillion per year (not counting fish farms or marine invertebrates).

Following is a sortable table of the world fisheries harvest for 2005. The tonnage from capture and aquaculture is listed by country. Capture includes fish, crustaceans, molluscs, etc. Countries which harvested less than 100,000 tons are not included.

Fish production in various countries in the world

Country	Capture	Aquaculture	Total
Algeria	126,259	368	126,627
Angola	240,000		240,000
Argentina	931,472	2,430	933,902
Australia	245,935	47,087	293,022
Bangladesh	1,333,866	882,091	2,215,957
Brazil	750,283	257,783	1,008,066
Cambodia	384,000	26,000	410,000
Canada	1,080,982	154,083	1,235,065
Chile	4,330,325	698,214	5,028,539
People's Republic of China	17,053,191	32,414,084	49,467,275

Contd...

Country	Capture	Aquaculture	Total
Colombia	121,000	60,072	181,072
Congo, Democratic Republic of the	220,000	2,965	222,965
Denmark	910,613	39,012	949,625
Ecuador	407,723	78,300	486,023
Egypt	349,553	539,748	889,301
France	574,358	258,435	832,793
Germany	285,668	44,685	330,353
Ghana	392,274	1,154	393,428
Greece	92,738	106,208	198,946
Greenland	216,302		216,302
Hong Kong	161,964	4,130	166,094
Iceland	1,661,031	8,256	1,669,287
India	3,481,136	2,837,751	6,318,887
Indonesia	4,381,260	1,197,109	5,578,369
Iran	410,558	117,354	527,912
Ireland	262,532	60,050	322,582
Italy	298,373	180,943	479,316
Japan	4,072,895	746,221	4,819,116
Kenya	148,124	1,047	149,171
North Korea	205,000	63,700	268,700
South Korea	1,639,069	436,232	2,075,301
Latvia	150,618	542	151,160
Lithuania	139,785	2,013	141,798
Madagascar	136,400	8,500	144,900
Malaysia	1,214,183	175,834	1,390,017
Mauritania	247,577		247,577
Mexico	1,304,830	117,514	1,422,344
Morocco	932,704	2,257	934,961
Myanmar	1,742,956	474,510	2,217,466
Namibia	552,695	50	552,745
Netherlands	549,208	68,175	617,383
New Zealand	535,394	105,301	640,695
Nigeria	523,182	56,355	579,537
Norway	2,392,934	656,636	3,049,570
Oman	150,571	173	150,744
Pakistan	434,473	80,622	515,095
Panama	214,737	8,019	222,756
Papua New Guinea	250,280		250,280
Peru	9,388,662	27,468	9,416,130
Philippines	2,246,352	557,251	2,803,603
Poland	156,247	36,607	192,854

Contd...

Country	Capture	Aquaculture	Total
Portugal	211,757	6,485	218,242
Russian Federation	3,190,946	114,752	3,305,698
Senegal	405,070	193	405,263
Seychelles	106,555	772	107,327
Sierra Leone	145,993		145,993
South Africa	817,608	3,142	820,750
Spain	848,803	221,927	1,070,730
Sri Lanka	161,960	1,724	163,684
Sweden	256,359	5,880	262,239
Taiwan(Republic of China)	1,017,243	304,756	1,321,999
Tanzania	347,800	11	
Thailand	2,599,387	1,144,011	3,743,398
Tunisia	109,117	2,665	111,782
Turkey	426,496	119,177	545,673
Uganda	416,758	10,817	427,575
Ukraine	244,943	28,745	273,688
UK	669,458	172,813	842,271
USA	4,888,621	471,958	5,360,579
Uruguay	125,906	47	125,953
Vanuatu	151,079	1	151,080
Venezuela	470,000		

Principles of Rationing

Principles of Rationing of Dairy Cattle

1. Dairy animals must be provided an ideal ration consisting of the following qualities:

    (a) Adequate amount of different nutrients. *(b)* Quite appetizing.

    (c) Easily digestible or palatable.

    (d) Produces desirable flavours to milk. *(e)* Contains varieties of feeds in it.

    (f) Consists of plenty of succulent and green feeds. *(g)* Economical.

    (h) Possibly fresh.

    (i) Properly balanced.

2. Providing such ration that has good effect on health.

3. Providing a laxative ration to keep normal digestion.

4. Feeding a ration, e.g., bulky ration to keep normal tone of paristaltic movement of the alimentary canal.

5. Providing the ration that is not toxic.

6. Dairy animals must be fed liberally.

7. Animals must be fed individually.

8. Feeding rations at regular scheduled intervals daily.

9. Avoid sudden change in rations of dairy animals.

10. To ensure maximum consumption of dry matter, the green fodder must be sufficiently provided.

11. Fodders like silage that have smell must be fed after milking so that off flavour may not be imparted to milk.

12. Dairy animals can consume total dry matter 2 to 3% of their live body weight. Hence total need of dry matter be accordingly determined.

13. Maximum quantity of green fodder can be given to cow varies from 35 to 40 kg/day depending upon feed quality, palatability and demand of animal.

14. Green fodders especially the improved varieties and legumes such as lucerne, berseem, and cowpea could be fed to dairy animals to replace concentrates @ 1 kg. conc. for 12 kg of greens.

15. Leguminous green fodder like berseem, lucerne, cowpea etc. must not be fed on empty stomach to dairy animals as this may upset the digestion and cause bloat. Always either the dry fodder like wheat bhusa is mixed or fed first and then greens.

16. Animals should not be overfed with concentrate for it would not be an economical policy.

17. Feeding hay must not be done just before and at milking time, for it may create dusty atmosphere in the barn and adversely affect quality of raw milk.

18. Most dairy farmers prefer feeding grains at milking time as this becomes part of stimulus for letdown of milk.

19. Concentrate provided to dairy cows must contain 15 to 175 DCP and 705 TDN.

20. Concentrate may be supplied to cows as follows:

	Maintenance	Production
Cows	1 kg	1 kg for every 3 kg milk
Buffaloes	1.5 kg	1 kg for every 2.5 kg milk

21. Dairy cows need large intake of water daily to meet their maintenance and production requirement, otherwise shortage of drinking water would

cause serious depression of milk yield. Therefore plentiful supply of clean water should be made available to dairy cows at least 3 times daily, if not at all times.

Factors affecting composition of milk

There is a great deal of variation in the composition of milk; even the composition of the same animal is not always the same. Amongst the milk constituents, the fat content of the milk is most variable. After fat the other constituents vary in the following order: Protein, lactose, ash. The factors responsible for such variations in the composition are given below:

1. ***Species of animal:*** Milk composition of different species of animals varies from each other. The table on the opposite page presents the composition of human milk and that of few species of animals.

2. ***Influence of breed:*** The milk of some breeds of same species is comparatively higher in fat content than those of other breeds, e.g., milk of Red Sindhi cow contains higher fat than those of Holstein and Brown Swiss.

3. ***Individual variations:*** Individuality of animals is responsible for some of the greatest variations in the composition of milk. It is partly a matter of inheritance. It is partly due to unknown factors that operate in the case of individual animals.

4. ***Stage of lactation:*** The period from the time the calf is born until the cow ceases to give milk is called the 'period of lactation'. The secretion of milk immediately after calving is known as colostrums. It may last from 3 to 6 days. It contains more protein and more total solids than those of normal milk. The composition of colostrum changes rapidly in successive milking.

5. ***Effect of age:*** There is a tendency for the fat content to rise from the 1st to 3rd lactation period, then remains fairly constant in subsequent lactation periods but later towards advancing age there is slight reduction in the fat content of the milk.

6. ***Seasonal variation:*** Generally fat content of milk tends to increase in cold weather and to decrease in warm weather. The solids-not-fat content also varies to some extent in similar manner.

7. ***Variations from milking to milking:*** The composition of the milk of an animal may show variation in different milking. Such variations are noted even when all other conditions are the same. The reasons for such variations are not yet known.

8. ***Length of interval between milking:*** When the milking is done twice in Z4 hours the milk drawn after longer interval contains less fat percentage

than that drawn after shorter intervals. The morning milk which is drawn after longer interval contains lesser fat as compared to that of evening milk.

9. **First and last milk:** There is a considerable variation in the fat content of the 'for milk' 'mid milk' and 'strippings'. The fore milk is very poor in fat and strippings are very rich in fat. This variation is more when the milk yield is high

10. **Feed of the animals:** When the animals are given sufficient balanced ration, feed has no significant effect on composition. When the feeding is changed there is some variation in the composition of milk but such variations are temporary.

11. **Physical condition of the animal:** There may be a change in the composition of milk if the animal is suffering from any disease viz., in the milk of cow or buffalo suffering from mastitis there is reduction in fat, protein and lactose content and marked increase in chlorides content.

12. **Weather conditions:** Weather conditions also create an impact on the composition of milk. During a long period of drought in summer, the yield and solids-not-fat content is reduced but the fat content is increased.

13. **Environment at the time of milking:** Anything which causes discontentment and uneasiness in the cow at the time of milking causes the cow to be nervous and leads to the holding up of her milk. As the last portion of the milk which is held is rich in fat, hence it affects the composition.

14. **Exercise:** There are many degrees of exercise.

    Light exercise often stimulates milk production without appreciable change in the percentage of fat. In the case of severe exercise, as when cows are used as work animals, the volume of milk is usually reduced but the fat content of milk is increased.

15. **Milker:** If the milker is not an efficient one and not able to draw milk completely from the udder, the fat content of milk is reduced because the last milk is very rich in fat.

TYPES OF MARKET MILK

Whole milk

Milk as such, derived from the animals without altering the composition, is termed as whole milk. When the milk is from a cow it is known as cow milk and that from a buffalo is known as buffalo milk. The minimum requirements of fat and S.N.F. for this milk have been fixed by P.E.A. for different States and

Union Territories. In most of the States cow milk is supposed to contain fat not less than 3.5% and S.N.F., not less than 8.5% whereas buffalo milk should contain fat not less than 6% and S.N.F. not less than 9%.

Standardized milk

This is milk that's fat and SNF contents have been adjusted to a certain pre-determined level. The standardization can be done by partially skimming the fat in the milk or admixing skim milk in proper proportions. Under the PFA Rules (197'6), the standardized milk contains a minimum of 4.5% fat and 8.5% SNF.

Toned milk

Toned milk refers to milk obtained by the addition of water and skim 'milk' powder to whole milk. Under that, PFA rules (1976), toned milk should contain a minimum of 3% fat and 8.5% SNF.

Double toned milk

It is similar to toned milk, except that under the PFA rules (1976) double toned milk should contain a minimum of 1.5% fat and 9% SNF.

Reconstituted milk

This refers to milk prepared by dispersing whole milk powder in water approximately in proportion of 1 part powder to 7-8 parts water.

Recombined milk

This refers to the product obtained when butter oil, skim milk powder and water are combined in the correct proportion to yield fluid milk. The milk fat may also be obtained from other sources like unsalted butter, fresh sweet cream or· plastic cream. Under the PFA rules recombined milk should contain a minimum of 3% fat and 8.5% SNF.

Filled milk

When milk fat in milk is substituted with some vegetable oil or fat, it is termed as filled milk.

CREAM

Cream may be defined as a fat-rich portion of milk or that portion of milk into which has been gathered and which contains a large portion of milk fat. According to the PFA Rules (1976) it should contain a minimum of 25% fat.

Cream may be separated from milk either by gravity or centrifugal method. The basic principle of cream separation, whether by gravity or centrifugal methods, is based on the difference in specific gravity of milk fat and milk serum. The average specific gravity of milk fat and serum at 16°C (60°F) is 0.93 and 1.036 respectively. When milk is subjected to either gravitational or centrifugal force, the two components, viz., cream and skim milk, by virtue of their differing specific gravities, stratify or separate from one another.

Factors affecting the fat content of cream in centrifugally separation

(a) **Position of cream screw:** The fat content of the cream is regulated by adjusting the cream screw. This screw can be driven in or out When the screw is moved in higher fat percentage cream is obtained and vice versa.

(b) **Richness of the milk :** Other conditions remaining the same, the cream produced from richer milk will contain more fat than the cream produced from milk of lesser fat content.

(c) **Speed of bowl :** The fat test of the cream is increased with higher speed and decreased with lower speed. With the increase in speed the centrifugal force generated is greater which is responsible for this change.

(d) **Rate of inflow of milk :** An increase in the rate of inflow of milk in the separator bowl causes a reduction in the percentage of fat in the cream and vice versa.

(e) **Temperature of milk :** An increase in the temperature of milk causes a decrease in the fat percentage in the cream and vice versa.

(f) **Amount of flushing :** The amount of skim milk or water used for flushing will affect the fat percentage. An excessive amount will cause the fat test of the cream to be lowered.

(g) **Formation of separator slime :** A reduction in the fat percentage of cream may result if a large quantity of milk is separated in a stretch without cleaning the bowl and removing the separator slime.

Types of Cream

1. **Artificial cream :** It is a food article made up of milk ingredients to which water has been added to make a product resembling naturally produced cream.

2. **Cereal cream :** A very light commercial sweet cream usually testing about 10 to 12% butter fat. It is more commonly known under the name of cereal milk but is sold in the market as "Half-n-half".

3. **Clotted cream :** Cream prepared by a process of scalding and cooling which results in a thick, heavy, clotted substance. It is used chiefly in place of butter and also known as Evaporated Cream.

4. **Coffee cream :** Cream used in Coffee. This cream is usually homogenized and tests about 20% butterfat. It is also called "Single Cream", "Table Cream" or "Homogenized Cream", if it has been homogenized. The following features have been suggested for desirable table or coffee cream.

   (a) Fat content-20-25%.

   (b) Maximum viscosity for a given fat percentage.

   (c) No "oiling off" in hot coffee.

   (d) No "feathering" in hot coffee.

   (e) No cream plug formation.

   (f) Minimum amount of serum of skim milk in bottom of bottle.

   (g) No visible sediments.

   (h) Pleasing flavour: no feed cooked or oxidized flavour.

   (i) No developed acidity.

   (j) Maximum ability to colour coffee.

   (k) Low bacterial count.

5. **Cultured cream :** A ripened cream of high acidity characterized by a clean flavour, smooth texture and very heavy body. The product has an acidity of about 0.6% to 0.8% which is derived from a commercial starter culture. It may be used as a dressing with cottage cheese, salads and fruits, as a sandwich spread and in various, other ways. After standardizing the cream to at least 18% fat, pasteurize at not less than 82.2°C for minimum 10 minutes and preferably 30 minutes. Homogenize at a pressure 2000 p.s. 1. Cool to 21. 1°C and add 1·2% starter which will ripen the product to an acidity of 0.6 to 0.8% in 12-16 hrs: Package with a minimum of agitation, cool to 4.5'C or lower and age for 12-24 hrs prior to marketing.

6. **Frozen cream :** Sweet cream which has been frozen and held at low temperature for later use in dairy manufacturing. Because ice-cream is largely consumed during hot weather and on special holidays many ice-cream manufacturers follow the practice of buying at low price during the surplus month of good quality cream and storing it as frozen cream until needed. Good quality fresh cream, which tests at least 40% fat and has been pasteurized, should be used. It is usually stored in straight side tins or 50 to 60 lbs capacity or in fiber containers and kept at temperature of from -18° to-23°C. These containers should have no exposed iron or copper.

7. **Heavy cream :** Sweet cream containing not less than 36% rat.

8. **Light cream :** Sweet cream containing 18-30% fat (The term sweet cream is used for the fresh cream which has an acidity of less than 0.2%).

9. **Plastic Cream :** A heavy concentrated cream containing about 80% fat. It is plastic in form, much the same as butter. In spite of its high fat concentration, the fat remains in substantially its original emulsion with the solids-not-fat present in the serum. Plastic cream is produced commercially by the use of a separator designed for skimming a very rich cream. Either whole sweet milk or sweet cream testing 40% fat or less may be used. After separating, the cream is quickly cooled and packaged in better tubs. Plastic cream is used in the manufacture of butter, ice-cream, coffee cream, cream cheese and other dairy products.

10. **Synthetic cream :** A product made of non-milk fats such as cotton seed oil, peanut oil, and certain ingredient of soyabean to which lecithin and other ingredients have been added. It is used as a substitute for cream.

11. **Whipping cream :** A grade of market cream especially adapted for whipping in the open air. Whipping cream should contain 32 to 40% fat and should be aged for at least 24 hours at a temperature from 4.5°-10°C to increase clumping of fat globules. Rapid agitation of whipping is important for best results.

Common defects of milk and cream

1. *Barny flavour :* A flavour suggestive of contamination of milk with manure and stale air. It may be due to the actual contamination of milk with manure or more likely to cows being housed in unventilated stables where the air is strongly tainted with barny odours. In such cases the cows breathe in the tainted air, the blood picks up the substances from the lungs and same find their way to the secreting cells of the udder. The flavour is actually secreted in the milk. Barny flavour is more common in cold weather when the stables are closed up for warmth. Lack of proper cleaning of stable is a factor.

2. *Bitter flavour :* An off-flavour of milk and milk products which may result from several different causes, the most common probably being bacterial action and the eating of certain feeds and bitter weeds by the cows. Milk from cows advanced in lactation is often bitter as a result of high lipase content and rancidity.

3. *Bloody milk :* Milk appearing red, especially near the bottom of the bottle or container. This is usually due to blood which comes into the milk from a ruptured blood vessel in the udder. It is generally caused by injury to the udder.

4. ***Blue milk :*** A very uncommon condition of milk caused by a bacterium, *Bacillus cyanogenes.* The blue colour is developed when milk is allowed to stand. Immediate pasteurization of the milk will prevent this condition from developing.

5. ***Cooked flavour :*** A flavour defect of dairy products which results from excessive high pasteurization temperatures or improper methods of heating. *Cream pi ug:* A defect of milk or cream characterized by a thick layer of buttery cream at the top so that the liquid does not from its container. It is caused by partial churning of fat globules. Therefore, all causes of excessive agitation and foaming should be avoided. Freezing is liable to cause this defect due to pressure exerted by the expanding ice crystals rupturing the fat globule membrane, causing them to stick together and form small butter particles.

6. ***Feathering :*** A defect of table cream indicating that it has not been properly processed. As a result, there is some coagulation of the cream in hot coffee, a condition in which the cream forms a flaky feathery condition at the top of the cup. It usually does with unhomogenized cream but with homogenized cream it is because of using excess pressure in homogenization.

7. ***Feed flavour :*** A very common defect of milk and cream. This defect is especially noticeable when cows are fed green feeds or succulent feeds such as silage, cabbage and turnip, just before milking. Even alfalfa, clover and grass when fed just prior to milking are apt to produce an objectionable flavour in milk. Garlic, wild onion, ragweed and leek are a few of the many weeds which cause feed flavour.

8. ***Foreign flavour :*** A term used to designate flavours not normal or natural to milk or the dairy products. The flavours of gasoline, tobacco, chlorine, medication, disinfectants etc. are classed as foreign flavours.

10. ***High acid flavour :*** A flavour defect of dairy products caused by the development of excessive amount of lactic acid in milk or cream which appears as a result of holding the product at too high temperature for too long period of time.

11. ***Malty flavour :*** An off-flavour of milk or cream. It is generally caused by the growth of a certain strain of lactic acid bacteria *streptococcus lactic* var maltigenes.

12. ***Metallic flavour :*** A common flavour defect of milk and other dairy products brought about by the individual cow, the action of copper and copper alloys in the equipment or direct sunlight. Metallic flavours may have their origin on the farm by exposure to rusty utensils or in milk plants or creameries by exposure to copper or iron surfaces, especially when the milk is warm. The oxidative offflavour which if continued

gives carbohydrates and tallow flavours, is even more objectionable, and usually follows metallic flavour.

13. **Oiling-off :** A cream defect caused by destabilization of fat emulsion forming drops of fat on the surface of coffee. Some reasons for oiling-off are; breaking of fat emulsion by partly frozen milk before separation; separation temperature above 32(mechanical agitation before separation; separation of cream testing 45% fat or more; holding cream warm for several hours before pasteurization; cooling cream in vats etc. This defect can be rectified by homogenization, avoiding unnecessary agitation at every process, heating rapidly and flash cooling, holding at 63°C for 30 minutes with slow agitation and holding cream at low temperature.

14. **Rancid flavour :** A rancid flavour results when milk fats are hydrolyzed by enzyme known as lipase. Raw milk usually contains lipased enzyme. Thus, if raw milk while warm is subjected to sufficient agitation to rupture some of the fat globule membranes, the lipase comes in contact with the fat and rancidity results. Pasteurization or a heat treatment will inactivate lipase enzyme.

15. **Ropy milk :** A type of abnormal milk characterized by a sliminess or ropiness varying in degree from a slightly increased viscosity to ropiness so pronounced that the milk may be drawn out in threads several feet long. This condition is due to growth of micro organisms.

CHEESE

Cheese as defined by Davis is a product made from curd obtained from milk by coagulating the casein with the help of rennet or similar enzymes in the presence of lactic acid produced by added or adventitious micro organisms, from which part of the moisture has been removed by cutting, cooking and/or pressing, which has been shaped in a mould, and then ripened by holding it for some time at suitable temperatures and humidity.

According to PFA Rules (1976) hard cheese shall contain no more than 43% moisture and not less than 42% milk fat in dry matter.

A number of varieties of cheeses are available in the market. These may be classified as follows:

Among all varieties Cheddar cheese is the most popular variety. It was developed in the village of Cheddar in Somersetshine, England, in the 16th century.

Method of manufacturing Cheddar cheese

Reception of milk

Only high grade milk should be accepted for manufacture of cheese. Usually cow milk is preferred for cheese; however buffalo milk can also be used for the same. Milk received at cheese factory should be carefully examined for its quality before accepting.

Filtration/clarification

The chief object of filtration/clarification is to remove any visible dirt in milk so as to improve the aesthetic quality of the cheese. The milk is usually preheated to 35-40°C for efficient filtration/clarification.

Standardization of milk

In cheese-making, standardization refers to adjustment of the fat and casein ratio in cheese milk to 1:0.7 (approximately). The chief object of standardization is to maintain fat and dry matter ratio of cheese in day-today preparations.

Pasteurization

Milk is pasteurized either by holding or H.T.S.T. method to destroy pathogens and fault-producing microorganisms. It also helps to produce a more uniform product of high quality. Besides these advantages there are certain limitations of pasteurizing cheese milk, like: (a) It destroys the typical flavour and body of cheese. (b) It entails a longer ripening period. (c) It encourages the use of low quality milk. (d) It increases the overall cost of cheese-making.

But the advantages of pasteurizing cheese milk heavily outweigh its limitations. After pasteurization cheese milk is put into the cheese vat where rest of the processing is carried out.

Addition of calcium chloride

During pasteurization of milk, part of soluble calcium in milk is rendered insoluble. This results in slower renneting action and a weaker curd. This loss of soluble calcium is compensated by the addition of 0.01 to 0.3% calcium chloride to milk.

Addition of starter

In cheese vat temperature of milk is brought to 30-31°C and cheese starter is added at a rate of 0.5 to 1%. A cheese starter usually contains *Streptococcus lac tis* and/or *S. cremoris.* Before adding, the starter should be stirred until smooth and creamy in consistency, then strained and added in the required quantity and mixed thoroughly and uniformly into the milk.

Addition of colour

Cheese colour is extracted from annatto seeds. It is soluble in water. It is added just before renneting: 30 to 200 ml colour per 1000 kg milk is added after diluting it with 20 parts of potable water for even distribution. It is vigorously agitated to ensure uniform mixing.

Addition of rennet

Rennet is the preparation obtained from the fourth stomach of the young calf. It contains two chief enzymes; viz., rennin and pepsin. Rennin is an extremely powerful clotting enzyme, which causes rapid coagulation without much proteolysis, whereas pepsin induces proteolosis, leading to bitterness in cheese Rennet is available as a liquid or powder or as tablets. Rennet is added to the milk after the acidity has been increased by 0.02% over initial acidity. Usually liquid rennet is added at a rate of 15 to 25 ml/100 liter of milk after diluting with 20 to 40 parts of water for uniform distribution. The milk is thoroughly stirred to ensure even mixing.

Setting the curd

After addition of rennet, vat is left undisturbed for setting the curd. The temperature should be maintained at 30-31°C by controlling the temperature in the outer jacket of cheese vat.

Cutting the curd

When the curd has been set firmly, it is curd not cubes of a specific size. Curd is cut with the help of horizontal and vertical cheese knives. It is first cut with horizontal knife by moving it lengthwise, then with vertical knife lengthwise, widthwise. Acidity of the curd at the time of cutting is almost equal to acidity of initial milk.

Cooking the curd

After cutting, the curd is left undisturbed for 2-3 minutes so that curd cubes can attain some firmness. Then gentle stirring is started. The speed of stirring increases with the gradual firming of curd cubes. The heat is applied slowly by increasing the temperature of water in outer jacket of cheese vat. The heating is gradually increased. First 1°C temperature is increased in 15 minutes and thereafter to the maximum cooking temperature (37 to 39°C) at the rate of 1°C every 4 minutes. During cooking the gentle stirring is continued to avoid matting of curd cubes. At the end of cooking the acidity of whey is raised to about 0.135%.

Draining of whey

When the curd cubes have attained enough firmness and has been reduced to about one-half of their size at cutting, stirring is stopped and whey is drained off from the vat.

Cheddaring

After draining of whey, the curd cubes are allowed to mat for 5 to 10 minutes. Matted curd is cut into small slabs. These slabs are turned upside down at regular intervals followed by piling and repiling one over another. This process usually lasts two hours or more. At the end of cheddaring, the curd becomes drier, more mellow and silky and resembles chicken breast meat. The acidity of whey reaches 0,45 to 0.55%.

Milling

It is a mechanical operation of cutting the curd blocks into small pieces with the help of a cheese mill. It promotes further expulsion of whey and enables quick distribution of salt in curd.

Salting

The common powdered salt at a rate of 1-2% by weight of green cheese is sprinkled over milled cheese and mixed well. Salting helps in further removal of whey, checking undesirable fermentation and improving the quality of cheese.

Hooping and pressing

The curd, after salting, is filled in cheese cloth· lined hoops. These hoops are then kept under the cheese press for pressing the cheese. The pressing is done in two stages. First the cheese is pressed for about half an hour by applying a maximum pressure of $0.5 \text{ kg}/\text{cm}^2$. Then the cheese block is removed from the hoop and dressed.

Final pressing is done for about 12-24 hours by applying a pressure of about $2 \text{ kg}/\text{cm}^2$.

Drying

After the pressing is over, the cheese blocks are removed from the hoops and kept in drying room for a few days to form rind all around the surface. In the drying room the temperature is maintained at 12 to 16°C and the relative humidity at 50%.

Paraffining

After the formation of rind, the blocks are dipped in a hot paraffin for a few seconds to apply a thin coating of paraffin. Paraffining reduces the loss of moisture during ripening, prevents extensive mould growth and protects the cheese against insects.

Ripening

Ripening or curing of cheese refers to the storage of cheese for at least 2 to 3 months in a curing room where a temperature of 10-16°C and 75 to 85% relative humidity is maintained. During ripening physical, chemical and bacteriological properties are profoundly changed and a characteristic flavour and body and texture are developed.

MILK POWDER

Milk powder or dried milk is the product obtained by the removal of water from milk by heat or other suitable means to produce solid containing 5% or less moisture. Whole milk, skim milk or partially skimmed milk may be used for the purpose of drying.

According to PFA Rules (1976) whole milk powder shall contain no more than 5% moisture and not less than 26% fat. Partially skim milk powder shall contain no more than 5% moisture and the fat content of the product shall be between 1.6 and 24%. Skim milk powder shall contain no more than 5% moisture and not more than 1.5% fat.

There following are the two systems of drying milk:

1. Drum drying or doller drying.
2. Spray drying.

DRUM DRYING

Principle

Concentrated milk is applied in a thin film upon the smooth surface of a continuously rotating steam-heated metal drum, roller or cylinder. The film of dried milk is continuously removed by a stationary knife/doctor's blade/scraper. The drums/ rollers are normally horizontal hollow cylinders 8-12 ft. in length and 2-4 ft in diameter. They are heated internally by steam. In case of double drum driers drums are mounted parallel and one drum is fixed on a movable frame so that the gap between the drums could be adjusted as desired. The speed of the drums is also adjustable. The average speed ranges between 12·20 rpm. The milk film is removed from the drum after nearly 3/4 of revolution of the drum has taken place. The product is in contact with the drum for about 3 sec. or less at a temperature of about 150°C.

Manufacturing process

The concentrate milk with about 18% total solids is pumped into the reservoir between the two drums of roller drying equipment. The operating conditions are adjusted to regulate the capacity of the drier and to control the moisture content in the milk powder. Under proper drying conditions a uniform thin sheet of dry milk is scraped from each drum. The film drops into a trough and a screw conveyer breaks the film into small pieces while transporting to the packaging station. Further reduction in particle size can be done in a hammer mill. The ground particles are sieved through a mesh and packed in 25 or 50 kg Kraft paper bags with a plastic liner in case of skim milk powder and gas packed in metal containers in case of whole milk powder.

SPRAY DRYING

Principle

The basic principle consists in atomizing the concentrated milk to form a spray of very minute droplets (fog-like mist) which are directed into large, suitable designed drying chamber where they mix intimately with a current of hot air. Owing to the large surface area the milk particles surrender their moisture practically instanteously and dry to a fine powder which is removed continuously.

Manufacturing process

Concentrated milk with 35 to 45% total solids is atomized either with pressure type or centrifugal type atomizer. The atomized droplets of milk are dried within a chamber with inlet hot air at 150°C – 230°C and outlet air at 75°C – 100°C depending upon the drier characteristics. To reduce heat damage during dehydration and yet obtain the desired moisture in the powder of low exhaust air temperature is preferred. The atomized product is brought into intimate contact with heated air in the drying chamber for moisture removal. As the product is dried, it is necessary to separate the dried product from the air. Cyclone collector is most commonly used for powder collection. Multiple cyclones of relatively smaller diameter increase the efficiency of powder recovery. The dried particles should be removed from the drying chamber as quickly as possible to reduce the heat damage to the product. The dried product collected from cyclone separator is cooled and packaged. Skim milk powder is packed in kraft paper bags with a plastic liner and whole milk powder is gas packed in metal containers. A mixture of nitrogen and carbon dioxide is used for gas packaging of whole milk powder.

Human milk is usually regarded as ideal food for the feeding of infants, that is, children below one year of age. Yet owing to various causes, an infant may not derive adequate nutrition from the mother and it may become necessary to feed the infant either wholly or partly with milk from other sources. Infant milk food has been formulated to meet the feeding requirements of infants who do not get sufficient feeding from their mother. Infant milk food refers to the material prepared either by spray drying or by roller drying of the milk of cow or buffalo or a mixture thereof. The milk may be modified by partial removal of fat and by the addition of different carbohydrates like sucrose, dextrose and dextrins, maltose and lactose; salts like phosphates and citrates; vitamins A, B group, C and D and minerals like calcium and iron.

The infant milk food should be white or white with a greenish tinge to light cream in colour, free from lumps, reasonably free from brown or black specks and should be uniform in appearance. It shall be free from starch, vegetable fat or any non-milk fat. It should be free from extraneous matter which may be harmful to human health.

5

REASONING

- **Factors affecting composition of milk**

  Species, Breed, Individuality, Interval of milking, Frequency of milking, Completeness of milking, Irregularity of milking, Day-to-day milking, Disease and abnormal conditions, Portion of milking, Stage of lactation, Yield of milk, Feeding, Season, Age of animal, Condition of cow at calving, Excitement, Administration of drugs and hormones etc.

- **Accurate Method to measure the mil**

  By weight measurement of milk is more accurate than the measurement by volume.

- **Types of milk available in the market**

  Sterilized milk, Homogenized milk, Soft-curd milk, Flavoured milk, Vitaminzed milk, Irradiated milk, Frozen milk, Concentrated milk, Fermented milk, Standardized milk, Reconstituted milk, Rehydrated milk, Recombined milk, Toned milk, Double toned milk, Humanized milk, Miscellaneous milks, condensed milk, dried milk etc.

- **Detail list of milk products**

  Khoa/Mawa, Gulabjamun, Kalajamun, Pantua, Lalmohan, Burfi, Kalakand, Milk cake, Peda, Dharwad peda, Thirattupal, Rabri, Khurchan, Basundi, Kulfi, Malai-Ka-Barafi, Falooda, Kunda, Bal mithai, Paneer, Chhana, Rasgolla, Rasomalai, Rajbhog, Khirmohan, Sandesh, Chhana-Murki, Cham-Cham, Chhana podo, Surti paneer, Bandel Cheese, Ksheer Sagar, Sita Bhog, Chhana Gaja, Chhana Jhele, Chhana Kheer, Chhana pakora, Rasaballi, Shosim, Kalari, Dahi, Yogurt, Mishti Doi, Shrikhand, Chakka powder, Shrikhand wadi, Lassi, Mattha, Kadhi, Raita, Dahi vada etc.

- **Importance of lactose in human diet**

  Lactose is the principal carbohydrate of milk and it is exclusively found in dairy foods. Lactose provides 30 per cent of the total milk calories. It facilitates absorption of calcium, magnesium, and manganese. In infants it helps in the net retention of phosphorus. In the gastrointestinal tract, lactose promotes the growth of beneficial lactic acid bacteria. It also helps in biosynthesis of vitamins such as biotin, riboflavin, folacin etc.

Thus, lactose helps to control the gastrointestinal disturbances and the accompanying health-related problems.

- **Dried milk products**

  Whey powder, cream powder, butter powder, ice cream mix powder, cheese powder, cheese powder, malted milk powder, dry sodium caseinate, srikhand powder, chhana powder, khoa powder etc.

DETECTION OF PRESERVATIVES IN MILK

Purpose

1. To check adulterants used as preservative in milk.
2. It is important to safe guard the quality of milk for making good dairy products.
3. It is also important to safeguard health of people if toxic substances are used as preservatives.
4. To make importance of pasteurization.

Procedure

Preservatives used for milk are as follows:

1. Boric acid and Borate   2. Carbonate and bicarbonate
3. Formalin   4. Hydrogen peroxide
5. B-Napthol   6. Salicylic acid
7. Benzoic acid   8. Fluoride
9. Colour annats

I. Boric Acid and Borate

Reagent

Conc. HCl, Turmeric powder, NH_4OH (28%)

Principle

Boric acid and borate give a characteristic red colour with turmeric paper forming a complex.

Procedure

Take about 5 ml milk sample in a test tube and add 1 ml conc. HCl and mix. Dip a piece of turmeric paper in acidified milk. Dry the paper and note the change of colour, if turns red then it shows presence of boric acid or its salts, but if paper is dipped in NH_4OH and turns green it indicates the presence of boric acid only.

II. CARBONATES AND BICARBONATES

Reagent

Alcohol (95) % Rosalie acid (1% alcoholic solution)

Principle

Rosalic acid develops arose red colour with milk containing carbonates and whereas it gives only a brownish colouration with pure milk.

Procedure

Take about 10 ml milk in a test tube, add 10 ml of alcohol and shake. Add 3 drops of rosalic acid and mix. Observe the colour. Rose red colour indicates the presence of carbonates and whereas brown colour indicates absence of carbonate.

III. FORMALIN

Reagent

$FeCl_3$ (%) conc., H_2SO_4. conc. HCl, $FeCl_3$ (10%)

Principle

Formaldehyde gives violet colour with ferric salts and other oxidizing agents.

Procedure: (Hehner Test)

(a) Take about 10 ml milk in a test tube and add 0.5 ml $FeCl_3$ (1%) add 5 ML H_2SO_4 very slowly by the side of test tube. Violet ring of the junction of two liquids indicate formalin.

(b) Now take 5 ml milk and add equal volume of solution (containing 1 ml of 10% $FeCl_3$ in 500 ml of conc. HCl solution) Heat for 5 minutes, Violet colour shows presence of formaldehyde.

IV. HYDROGEN PEROXIDE (H₂O₂)

Reagent

2% aqueous P-phenylene-diamine-hydrochloride.

Principle

The above reagent when used fresh, gives an intense blue colour with H_2O_2 Enzyme peroxidase liberates nascent oxygen which gives blue colour with indicator.

Procedure

Take 10 ml milk in a tube and add 2 drops of the reagent and mix. Observe the colour. Intense blue colour indicates the presence of H_2O_2.

Test 2. Warther's Test

Take 1 ml 1% Sodium Orthovandate and add 1 ml of 10% H_2SO_4 and mix. Add this mixture to 10 ml milk sample. Red colour shows presence of H_2O_2.

V. O² NAPTHOL

Make extraction of milk through chloroform. Separate the layer of chloroform and add drop by drop caustic potash solution. Dark green colour if develops, it shows presence of O²⁻ Napthol.

VI. DETECTION OF BENZOIC AND SALICYLIC ACID

Take 10 to 20 ml milk and equal volume of HCl and shake until curd gets dissolved.

Follow it to cool and add 25 ml ether and 25 ml petroleum ether.

Now add few drops of NH_4H in ether layer. If precipitate is observed it shows presence of benzoic acid.

Now add few drops of distilled water on ether layer and separate this ether layer from liter. Add few drops of neutral ferric chloride solution. If violet colour appears it indicates salicylic acid and grey precipitate shows benzoic acid.

Note: In above test the separate water layer obtained if treated with Bromine water boundary yellow ppt. shows salicicylic acid.

VII. Fluorides

Take 5 ml milk and add few drops of H_2O_2 mix. Now prepare reagent (titanium Sulphate + 10% H_2SO_4 in equal parts) and add 1 ml to milk sample. Presence of fluoride indicates almonds colour (brownish).

DETECTION OF ADULTERANTS IN MILK

1. **Tests for Annato :** Add acetic acid to 15 ml milk to set it as curd. Filter properly. Take curd and mix with ether in a conical flask, shake to mix it. Make extraction of fat by hoxhlet apparatus. Remove ether by evaporation. Collect residue in flask and add few drops of 0.1 N NaOH to make it alkaline. Filter it through paper. Wash the fat by water and separate out. The presence of orange brown stain on paper indicates presence of annatto. Put a drop of stenous chloride on paper then it gives colour showing presence of annatto.

2. **Presence of gelatin:** Take 10 milk and 10 ml acetic mercuric nitrate + 20 ml water shake the mixture and allow standing for 5 min. filter it through paper. Take little filtrate 3 ml and add 5 ml picric acid. Presence of yellow precipitate shows gelatin present.

3. **Cane Sugar:** Take 10 ml milk in a test tube and add 1 ml conc. HCl and about 0.1 g Resorcinol powder and mix. Place the tube in boiling water bath 5 min. Red colour indicates the presence of sucrose or cane sugar.

4. **Starch or Cereal Flours:** Take about 3 ml milk in a test tube. Boil with about 10 ml water. Cool and add few drops of iodine solution. If blue colour appears (on blue colour disappears), it shows milk is adulterated by cereal flour or starch.

5. **Skim milk Powder:** Take 5 ml milk sample in two centrifuge tubes and balance properly in centrifuge. Run it at 3000 rpm for 30 minutes. Decant the supernatant carefully. Dissolve the settled residue in 2.5 ml conc. HNO_3. Dilute with about 5 ml water and add 2.5 ml liquid ammonia. If orange colour develops, it shows presence of skim powder.

6. **Detection of Sodium Chloride:** Take 2 ml of milk and add 0.1 ml of 5 Potassium Chromate and 2 ml of 0.1 N silver nitrate. Appearance of red precipitate in the presence of sodium chloride.

7. **Detection of Urea in Milk:** Take 2 ml of milk and add 2 ml of p-Dimethyl Benzaldehyde reagent (1.6% of ethyl alcohol containing 10% HCl). Development of yellow colour denotes the presence of urea. The pure milk samples show a faint pink which should be ignored due to the presence of natural urea (up to 50 mg/100 ml) should be carried out with the control samples. A simple paper strip method has also developed using the above principle.

8. **Detection of Saccharin**: Curdle an aliquot of the diluted sample (about 25 ml) with dilute acetic acid. Shake well and filter. Acidify the clear filtrate with 2 ml of conc. hydrochloric acid and extract with 25 ml portion of ether. Draw of adequate layers and the combined ether extract with three successive portions of 5 ml of water. Evaporate the ether extract on water bath and add a drop or two of water mix well with glass rod and taste little. Characteristic sweet taste indicates the presence of the saccharin.

9. **Detection of Buffalo Milk in Cow Milk**: The presence of buffalo milk in cow milk can be detected by Hansa test, which is based on immunological assay. A drop of suspected milk after dilution with water (1:4) is treated with a drop of antiserum obtained by injecting buffalo milk proteins into rabbits. The characteristic precipitation reaction indicates the presence of buffalo milk.

10. **Detection of Skim Milk Powder in Milk**: Skim milk powder added to milk for adulteration gives an orange colour with nitric acid while milk without the powder gives a yellow colour.

11. **Detection of Added Colour**: The chief colouring materials which are considered here are some natural colouring material like Annatto, turmeric or coal-tar dyes. Some of these dyes are permitted only in some products. While the use of Annatto is prohibited in milk, its use is permitted in butter. To detect Annatto the milk fat is shaken with 2 per cent sodium hydroxide and the mixture is poured on filter paper. The filter paper absorbs the colour, which mains even after washing with water. When the stain is treated with a drop of 40 per cent Cl_2 and dried, a purple colour indicates the presence of Annatto.

Turmeric is detected when the colour, aqueous of alkali, extracted is treated with HCl. The resulting orange colour is treated with H_3BO_3 crystals; a red colour indicates the presence of turmeric.

Coaltar dyes adhere to animal fibers more firmly than natural colour. The curd of pure milk is white when extracted with ether but one containing coaltar dyes remains orange or yellow; this when treated with concentrated hydrochloric acid becomes pink.

12. **Detection of Pulverized Soap**: Soaps are generally defined as sodium and potassium salt of fatty acid. Therefore, to detect the presence of pulverized soap, Iodine Value refractive Index, fatty acid composition, salt ratio and ash content are excellent methods. The presence can also be judged by qualitative method. For example in 10 ml of milk, 10 ml hot water is added followed by 1-2 drops of phenolphthalein indicator solution. Development of milk colour indicates the presence of soap in milk.

13. **Detection of Vegetable Fat**: The adulteration of vegetable fat in milk can easily detected by the following methods. In case of synthetic milk, the fat is extracted either by Bose-Gottleib method or the fat extracted in butyrometer can also be used.

   (i) *Fatty Acid Composition:* Milk fat is characterized by lower chain fatty acids. For sample butric, capric, caprylic etc. whereas most of the vegetable fat do not contain these fatty acids. Therefore the adulteration of the vegetable fat can easily be detected by analyzing fatty acid profile by Gas liquid Chromatography.

   (ii) *Detection by Measuring Different Physico-chemical Properties:* The adulteration of vegetable fat can also be detected by measuring various physico-chemical properties. For example Refractive Index, RM and Polenske values, Iodine value etc.

   (iii) Hydrogenated vegetable oils like vanaspati are a common adulterant in milk fat. Its presence in milk fat can be detected by the fact that sesame oil is added in vanaspati as per the law. The presence of sesame oil can be tested by Baudoin test.

6

DISTINGUISH BETWEEN

Difference in measurement of milk by weight vs volume

Weight method	Volume method
1. Gives accurate reading, regardless of foam or temperature.	1. Not so accurate, as affected by foam and temperature, both influencing density.
2. Involves considerable initial expense for both apparatus and its installation.	2. Lower initial expenses.
3. Involves problems with maintenance.	3. Presents maintenance problems.
	4. Definitely a factor to be considered in the overall picture of sanitation.

Physico-chemical differences in synthetic MDK and Natural MUK

Characteristics	Synthetic Milk	Natural Milk
Colour	White	White
Taste	Extremely bitter	Palatable
Odour	Soapy becomes distinct on boiling	Characteristic milky odour
Texture	When rubbed between fingers gives a soap feeling	No soapy feeling when rubbed between fingers
pH	Alkaline. 9.0-10.5	6.6-6.8
Urea test	Distinctly positive	Very faint due to natural urea in milk
Sugar test	Positive	Negative
Neutralizer test	Positive	Negative
Vegetable fat test	Positive	Negative

REFERENCES

Sukumar, D.E. 1995. *Outline of Dairy Technology.* Oxford University Press, Delhi. pp. 539.

Aneja, R.P., Mathur, B.N., Chandan, R.C. and Baneree, A.K. 2002. *Technology of Indian Milk Products.* A Dairy Publication, Delhi. pp. 462.

Miller, G.D., Jarvis, J.K. and McBean, L.D. 2000. *Handbook of dairy foods and nutrition,* 2nd Ed. CRC Press, Boca Raton, Fl. USA. pp. 354.

Wesley, G.D. and May, W. 1980. *Animal Cell Culture Methods.* Blackwell Scientific Publishers. Edinburgh, UK.

Freshney, R. 1986. *Animal Cell Culture: A practical Approach.* E.S. Livingstone Ltd. Edinburgh, UK.

Gupta, P.K. 2000. *Elements of Biotechnology.* Rastogi Publications, Meerut, India.

Singh, B.D. 2000. *Biotechnology.* Kalyani Publishers, New Delhi.

Gupta, M.L. and Jangir, M.L. 2002. *Cell Biology: Fundamentals and Applications.* Agribios, Jodhpur, India.

Wiseman, A. 1988. *Principles of Biochemistry.* Survey University Press, New Delhi.

Roges, A. 1989. *Food Biotechnology.* Elsevier Applied Sci. Pub. London, UK.

Goldberg, I. 1994. *Functional Foods.* Chapman and Hall, New Delhi.

Chirkjian, J.G. 1995. *Biotechnology-Theory and Techniques.* Johnes and Barlet Publishers, London.

Byong, H.L. 1996. *Fundamentals of Food Biotechnology.* VCH Publishers, New York.

Sybenga, J. 1972. *General Cytogenetics.* North Holland Publishing Co., Amsterdam.

Bostock, C.J. and Summer, A.K. 1980. *The Eukaryotic Chromosomes.* North Holland Publishing Co., Amsterdam.

Sharma, A. and Sharma, A.K. 1980. *Chromosome Techniques.* Butterworths, London.

Maclean, N., Gregory, S.P. and Hanell, R.A. 1983. *Eukaryotic Genes.* Butterworths, London.

Warr, J. Rober 1985. *Genetic Engineering in Higher Organisms.* Edward Arnold Pvt. Ltd., Victoria, Australia.

About Indian Fisheries. National Fisheries Development Board - A Government of India sponsored autonomous organization. 2008.

Activities of NFDB. National Fisheries Development Board – Govt. of India. 2008.

Annual Report: India, 2008-2009. Department of Animal Husbandry Dairying and Fisheries, Ministry of Agriculture, Government of India. 2009.

Appleby, M.C.; J.A. Mench and B.O. Hughes (2004). *Poultry Behaviour and Welfare.* Wallingford and Cambridge MA: CABI Publishing. *ISBN 0-85199-667-1.*

Boadi, D. 2004. Mitigation strategies to reduce enteric methane emissions from dairy cows: Update review. Can. J. Anim. Sci. 84: 319-335.

Bollongino, Ruth and al. Molecular Biology and Evolution. *"Modern Taurine Cattle descended from small number of Near-Eastern founders"*. 7 Mar 2012. Accessed 2 Apr 2012. Op. cit. in Wilkins, Alasdair. io9.com. "DNA reveals that cows were almost impossible to domesticate". 28 Mar 2012. Accessed 2 Apr 2012.

Bradford, S.A., E. Segal, W. Zheng, Q. Wang, and S.R. Hutchins. 2008. *Reuse of concentrated animal feeding operation wastewater on agricultural lands*. J. Env. Qual. 37 (supplement) S97-S115.

Breward, J., (1984). *Cutaneous nociceptors in the chicken beak*. Proceedings of the Journal of Physiology, London 346: 56

Breward, J., (1985). *An Electrophysiological Investigation of the Effects of Beak Trimming in the Domestic Fowl (*Gallus gallus domesticus*)*. Ph.D. thesis, University of Edinburgh.

Breward, L. and Gentle, M.J., (1985). *Neuroma formation and abnormal afferent nerve discharges after partial break amputation (beak trimming) in poultry*. Experientia, 41: 1132-1134. doi:10.1007/BF01951693

Brown, David (2009-04-23). *"Scientists Unravel Genome of the Cow"*. The Washington Post. Retrieved 2009-04-23.

Capper, J.L. 2011. *The environmental impact of beef production in the United States: 1977 compared with 2007*. J. Anim. Sci. 89: 4249-4261

Clean Water Act (CWA) Concentrated Animal Feeding Operations National Enforcement Initiative". Epa.gov. Retrieved 2013-10-15.

Coupe, Sheena (ed.), Frontier Country, Vol. 1, Weldon Russell Publishing, Willoughby, 1989, ISBN 1-875202-01-3

Delbridge, A, *et.al.*, Macquarie Dictionary, The Book Printer, Australia, 1991

Delbridge, Arthur, The Macquarie Dictionary, 2nd ed., Macquarie Library, North Ryde, 1991

Devor, M. and Rappaport, Z.H., (1990). *Pain Syndromes in Neurology.*, edited by H.L. Fields, Butterworths, London, p. 47.

Dlugokencky, E.J. *et.al.*, 2011. *Global atmospheric methane: budget, changes and dangers*. Phil. Trans. Royal Soc. 369: 2058–2072.

Duncan I.J.H., Slee G.S., Seawright E. and Breward J., (1989). *Behavioural consequences of partial beak amputation (beak trimming) in poultry*. British Poultry Science, 30: 479–488

E.O. Wilson, The Future of Life, 2003, Vintage Books, 256 pages ISBN 067976811

Eckard, R. J. et al. 2010. *Options for the abatement of methane and nitrous oxide from ruminant production: A review*. Livestock Science 130: 47-56.

Edward O. Wilson, The Future of Life, 2003, Vintage Books, 256 pages ISBN 0-679-76811-4.

EPA. 2001. *Environmental and economic benefit analysis of proposed revisions to the National Pollutant Discharge Elimination System Regulation and the effluent guidelines for concentrated animal feeding operations*. US Environmental Protection Agency. EPA-821-R-01-002. 157 pp.

Export of marine products from India. Central Institute of Fisheries Technology, India. 2008.

Extremely drug resistant tuberculosis – is there hope for a cure? TB Alert – the UK's National Tuberculosis Charity. Retrieved 2007-04-02.

FAO. 1997. 1996 *Production Yearbook*. Food Agr. Organ., UN. Rome, Italy.

Fisheries. Tamil Nadu Agricultural University, Coimbatore. 2007.

Fishery and Aquaculture Country Profiles: India. *Food and Agriculture Organization of the United Nations.* 2011.

Friend, John B., Cattle of the World, Blandford Press, Dorset, 1978

Gentle M.J., Hughes B.O. and Hubrecht R.C., (1982). *The effect of beak-trimming on food-intake, feeding behaviour and body weight in adult hens.* Applied Animal Ethology, 8: 147–157

Gentle M.J., Hunter L.N. and Waddington D., (1991). *The onset of pain related behaviours following partial beak amputation in the chicken.* Neuroscience Letters, 128: 113–116

Gentle, M.J., (1986). *Beak trimming in poultry.* World's Poultry Science Journal', 42: 268-275

Gentle, M.J., (1992). *Pain in birds.* Animal Welfare, 1: 235-247

Gentle, M.J., Hughes, B.O., Fox, A. and Waddington, D., (1997). *Behavioural and anatomical consequences of two beak trimming methods in 1- and 10-d-old domestic chicks.* British Poultry Science, 38: 453-463

Gill, Victoria (2009-04-23). "BBC: Cow genome 'to transform farming'". BBC News.

Groves, C.P., 1981. *Systematic relationships in the Bovini (Artiodactyla, Bovidae).* Zeitschrift für Zoologische Systematik und Evolutionsforschung, 4:264-278. quoted in Grubb, P.(2005). "Genus Bison". In Wilson, D.E.; Reeder, D.M. Mammal Species of the World (3rd ed.). Johns Hopkins University Press. pp. 637–722. ISBN 978-0-8018-8221-0. OCLC 62265494.

Grubb, P. (2005). "*Bos taurus primigenius*". In Wilson, D.E.; Reeder, D.M. Mammal Species of the World (3rd ed.). Johns Hopkins University Press. pp. 637–722. ISBN 978-0-8018-8221-0. OCLC 62265494.

Harper, Douglas (2001). "*Capital*". *Online Etymological Dictionary.* Retrieved 2007-06-13.

Harper, Douglas (2001). "*Cattle*". *Online Etymological Dictionary.* Retrieved 2007-06-13.

Harper, Douglas (2001). "*Chattel*". *Online Etymological Dictionary.* Retrieved 2007-06-13.

Harvey, Fiona (17 May 2011). *Easing of farming regulations could allow milk from TB-infected cattle into food chain.*

Hui, ed. Ramesh C. Chandan, associate editors, Charles H. White, Arun Kilara, Y. H. (2006). *Manufacturing yogurt and fermented milks* (1. ed. ed.). Ames (Iowa): Blackwell. p. 364. ISBN 9780813823041.

India - National Fishery Sector Overview. *Food and Agriculture Organization of the United Nations.* 2006.

Injured cow in Nepal is serious matter. Yahoo! News. Retrieved 27 November 2010.

IPCC. 2001. Third Assessment Report. *Intergovernmental Panel on Climate Change.* Working Group I: The Scientific Basis. Table 4.2

IPCC. 2007. Fourth Assessment Report. *Intergovernmental Panel on Climate Change.*

IPCC. 2007. Fourth Assessment Report. *Intergovernmental Panel on Climate Change.*

Jacobs, G. H., J. F.Deegan, and J. Neitz. 1998. *Photopigment basis for dichromatic color vision in cows, goats and sheep.* Vis. Neurosci. 15:581–584

Kane, J.; Anzovin, S., & Podell, J. (1997). Famous First Facts. New York, NY: H. W. Wilson Company. p. 5. ISBN 0-8242-0930-3.

Krebs JR, Anderson T, Clutton-Brock WT, et al. (1997). *Bovine tuberculosis in cattle and badgers: an independent scientific review (PDF).* Ministry of Agriculture, Fisheries and Food. Retrieved 2006-09-04.

Kuenzel, W.J. (2001). *Neurobiological basis of sensory perception: welfare implications of beak trimming.* Poultry Science, 86: 1273-1282

Lott, Dale F.; Hart, Benjamin L. (October 1979). *Applied ethology in a nomadic cattle culture.* Applied Animal Ethology (Elsevier B.V.) 5 (4): 309–319. doi: 10.1016 / 0304 – 3762 (79) 90102-0.

Lowry, C.A.; Hollis, J.H.; De Vries, A.; Pan, B.; Brunet, L.R.; Hunt, J.R.F.; Paton, J.F.R.; Van Kampen, E. et al. (2007). *Identification of an immune-responsive mesolimbocortical serotonergic system: Potential role in regulation of emotional behavior".* Neuroscience 146 (2): 756–772.

Lunam, C.A., Glatz, P.C. and Hsu, Y-J., (1996*). The absence of neuromas in beaks of adult hens after conservative trimming at hatch.* Australian Veterinary Journal, 74: 46-49

Madden, Thomas (May 1992). "Akabeko". OUTLOOK. Online copy accessed 18 January 2007.

Mahabharata, Book 13-Anusasana Parva, Section LXXVI. Sacred-texts.com. Retrieved 2013-10-15.

Martin, C. et al. 2010. *Methane mitigation in ruminants: from microbe to the farm scale.* Animal 4 : pp 351-365.

McDonald, J. M. *et al.,* 2009. Manure use for fertilizer and for energy. Report to Congress. USDA, AP-037. 53pp.

McIntosh, E., The Concise Oxford Dictionary of Current English, Clarendon Press, 1967

McWhirter, Norris & Ross, Guinness Book of Records, Redwood Press, Trowbridge, 1968

Meat & Livestock Australia, Feedback, June/July 2008

Meunier-Goddik, L. (2004). Sour Cream and Creme Fraiche. *Handbook of Food and Beverage Fermentation Technology.* CRC Press.doi:10.1201/9780203913550. ISBN 978-0-8247-4780-0.

Muruvimi, F. and J. Ellis-Jones. 1999. *A farming systems approach to improving draft animal power in Sub-Saharan Africa.* In: Starkey, P. and P. Kaumbutho. 1999. Meeting the challenges of animal traction. Intermediate Technology Publications, London. pp. 10-19.

National Aquaculture Sector Overview: India. *Food and Agriculture Organization of the United Nations.* 2009.

Nicholson, C.F., R.W. Blake, R.S. Reid and J. Schelhas. 2001. *Environmental impacts of livestock in the developing world.* Environment 43(2): 7-17.

Perception of Color by Cattle and its Influence on Behavior C.J.C. Phillips and C.A. Lomas†2 J. Dairy Sci. 84:807–813

Phaniraja, K.L. and H.H. Panchasara. 2009. Indian draught animals power. Veterinary World 2:404-407.

Shapouri, H. et al. 2002. The energy balance of corn ethanol: an update. USDA Agricultural Economic Report 814.

Sherwin, C.M., Richards, G.J and Nicol, C.J. 2010. A comparison of the welfare of layer hens in four housing systems in the UK. British Poultry Science, 51(4): 488-499.

Steinfeld, H. *et.al.,* 2006, Livestock's Long Shadow: Environmental Issues and Options. Livestock, Environment and Development, FAO.

Strassman, B.I. 1987. *Effects of cattle grazing and haying on wildlife conservation at National Wildlife Refuges in the United States.* Environmental Mgt. 11: 35-44.

Subramanian Swamy (21 November 2001). *"Save the cow, save earth".* Express Buzz. Retrieved 27 November 2010.

Takeda, Kumiko; et.al., (April 2004). *Mitochondrial DNA analysis of Nepalese domestic dwarf cattle Lulu.* Animal Science Journal (Blackwell Publishing) 75 (2): 103–110.doi:10.1111/ j.1740-0929.2004.00163.x.

The state of world fisheries and acquaculture, 2010. FAO of the United Nations. 2010.

US Code of Federal Regulations 40 CFR 122.23

US Code of Federal Regulations 40 CFR 122.23, 40 CFR 122.42

US EPA. 2012. Inventory of U.S. greenhouse gase emissions and sinks: 1990–2010. US. Environmental Protection Agency. EPA 430-R-12-001. Section 6.2.

USDA 1994. Agricultural Statistics 1994. US Government Printing Office, Washington. 485 pp. Table 377.

USDA. 2011. Agricultural Statistics 2011. US Government Printing Office, Washington. 509 pp. Table 7.6.

USDA. 2011. Agricultural Statistics 2011. US Government Printing Office, Washington. 509 pp. Table 7.1.

Van Vuure, C.T. 2003. De Oeros – Het spoor terug (in Dutch), Cis van Vuure, Wageningen University and Research Centrum: quoted by The Extinction Website: Bos primigenius primigenius.

Whyte, Mariam; Lin, Yong Jui (2010). *Bangladesh.* New York: Marshall Cavendish Benchmark. p. 144.ISBN 9780761444756.

Zeder, Melinda A. ed. (2006). *Documenting Domestication: New Genetic and Archaeological Paradigms.* University of California Press. p. 264. ISBN 0-520-24638-1.

www.ingramcontent.com/pod-product-compliance
Lightning Source LLC
Chambersburg PA
CBHW021434180326
41458CB00001B/274